****Hope for Cynics****

Hope for Cynics

The Surprising Science of Human Goodness

JAMIL ZAKI

GRAND CENTRAL

NEW YORK BOSTON

Copyright © 2024 by Jamil Zaki

Cover design by Jim Datz. Cover images by Shutterstock.
Cover copyright © 2024 by Hachette Book Group, Inc.

Hachette Book Group supports the right to free expression and the value of copyright. The purpose of copyright is to encourage writers and artists to produce the creative works that enrich our culture.

The scanning, uploading, and distribution of this book without permission is a theft of the author's intellectual property. If you would like permission to use material from the book (other than for review purposes), please contact permissions@hbgusa.com. Thank you for your support of the author's rights.

Grand Central Publishing
Hachette Book Group
1290 Avenue of the Americas, New York, NY 10104
grandcentralpublishing.com
@grandcentralpub

First Edition: September 2024

Grand Central Publishing is a division of Hachette Book Group, Inc. The Grand Central Publishing name and logo is a registered trademark of Hachette Book Group, Inc.

The publisher is not responsible for websites (or their content) that are not owned by the publisher.

The Hachette Speakers Bureau provides a wide range of authors for speaking events. To find out more, go to hachettespeakersbureau.com or email HachetteSpeakers@hbgusa.com.

Grand Central Publishing books may be purchased in bulk for business, educational, or promotional use. For information, please contact your local bookseller or the Hachette Book Group Special Markets Department at special.markets@hbgusa.com.

Library of Congress Cataloging-in-Publication Data

Names: Zaki, Jamil, 1980- author.
Title: Hope for cynics : the surprising science of human goodness / Jamil Zaki.
Description: New York : GCP, [2024] | Includes index.
Identifiers: LCCN 2023057409 | ISBN 9781538743065 (hardcover) | ISBN 9781538743089 (ebook)
Subjects: LCSH: Cynicism—Social aspects. | Social justice—Psychological aspects. | Hope.
Classification: LCC HM1011 .Z35 2024 | DDC 149—dc23/eng/20240226
LC record available at https://lccn.loc.gov/2023057409

ISBNs: 9781538743065 (hardcover), 9781538743089 (ebook)

Printed in the United States of America

LSC-C

Printing 1, 2024

For Luisa and Alma

Contents

Introduction		1

Section I: UNLEARNING CYNICISM

Chapter 1.	Signs and Symptoms	15
Chapter 2.	The Surprising Wisdom of Hope	29
Chapter 3.	Preexisting Conditions	46
Chapter 4.	Hell Isn't Other People	65
Chapter 5.	Escaping the Cynicism Trap	79

Section II: REDISCOVERING ONE ANOTHER

Chapter 6.	The (Social) Water Is Just Fine	97
Chapter 7.	Building Cultures of Trust	114
Chapter 8.	The Fault in Our Fault Lines	130

Section III: THE FUTURE OF HOPE

Chapter 9.	Building the World We Want	151
Chapter 10.	The Optimism of Activism	168
Chapter 11.	Our Common Fate	185

Epilogue	200
Acknowledgments	205
Appendix A: A Practical Guide to Hopeful Skepticism	209
Appendix B: Evaluating the Evidence	213
Chapter Claim Ratings	217
Notes	225
Index	263

Hope is not a lottery ticket you can sit on the sofa and clutch, feeling lucky... [It] is an ax you break down doors with in an emergency.

—Rebecca Solnit

Hope for Cynics

I was always jealous of Emile Bruneau. We were both professors of psychology. We both used brain science to study human connection, and hoped our work could help people connect more effectively. We presented at many of the same conferences and snuck out for martinis at many hotel bars, becoming fast friends along the way.

Emile probably made a lot of people jealous. A square-jawed ex–rugby player, he turned heads anywhere he went, which was everywhere he could. Emile worked to promote peace in Northern Ireland, biked across South Africa, and grappled with a local wrestling champion in Mongolia. At home, he assembled a Ford Model A, tended bees, and built his kids a tree house more elaborate than some New York City apartments. His professional accomplishments were just as impressive: Emile founded the Peace and Conflict Neuroscience Lab at the University of Pennsylvania, which pioneered scientific tools for overcoming hatred.

Emile was larger than life, but the thing I most envied about him was his hope. This might seem strange given what I do for a living. For two decades, I've studied kindness and empathy, teaching people around the world about the importance of these virtues. This has made me an unofficial ambassador for humanity's better angels, often recruited to jump-start people's faith in one another.

But all this time, I've lived with a secret. In private, I'm a cynic, prone to seeing the worst in people. This tendency began early; a chaotic family life made it hard for me to trust people's intentions. Since then, I've found stronger emotional footing through new relationships and been uplifted by science as well. My lab and I have discovered that most people value compassion over selfishness, that donating money activates similar parts of your brain as eating chocolate, and that helping others through their stress soothes our own. The message of our work is simple: There is good *in* us, and it does good *for* us.

But there's a difference between understanding something and feeling it. I've met miserable happiness experts and stressed-out meditation researchers. Scientists are sometimes drawn to what they have trouble finding in their own lives. Perhaps I've spent all this time charting a map of human goodness in the hopes of locating it more easily on the ground.

Recently, finding the good in others has become harder instead. Emile and I met in 2010. In the decade that followed, division, inequality, depression, and sea levels all rose. In my own social circle, I witnessed industrious, brilliant friends struggle to find work, let alone any semblance of the American dream. I joined Twitter to follow other scientists but encountered a deluge of outrage, lies, and personal branding. California caught fire, and the hilly vineyard where my wife and I had eloped was consumed. On our anniversary, we drove through its charred remains, wondering how much more of the world would look like this, how soon. I could recite evidence about kindness from my lab and a dozen others, but as the world seemed to grow greedier and more hostile, my instincts refused to follow the science.

Emile was one of the few people with whom I shared this struggle. Across many conversations, he tried to resuscitate my hope. Our science could teach people about the good inside them, he'd say, and about the fears that keep that good covered, like the sun behind clouds. We could move people toward community and justice—their true values.

Emile's plucky monologues seemed ridiculous, and sometimes made me wonder if we had that much in common after all. He had witnessed hatred on five continents. Where did he get off being so optimistic? His positivity seemed like wishful thinking, or the sign of a sheltered mind.

Then one day we talked about his childhood, and it became clear how wrong I'd been. Shortly after Emile was born, his mother was plagued by cruel, mocking voices, as inescapable to her as they were imperceptible to everyone else. She had developed severe schizophrenia and for the rest of her life she remained at war with her own mind, unable to raise Emile.

Yet when they were together, she protected him from the mayhem inside her. "She never let any of the darkness touch me," he remembered. "Even when she was in the pits of despair, she only gave me light." Hearing him recount this story, I realized Emile was the opposite of naive. He had seen firsthand how care could bloom in the face of immense pain. In fighting for our best side, he didn't have the luxury of seeing only our worst. His hope was like his mother's tenderness: a defiant choice.

In 2018, that hope would be tested again. His laptop screen looked dimmer each night; then the headaches began. As a neuroscientist, he recognized the warning signs, requested a CT scan, and discovered the brain cancer that would take his life two years later, at the age of forty-seven. The tragedy hit him and his family squarely. Emile's children, four and six years old, would grow up without their father. His wife, Stephanie, would lose her beloved partner. Decades of work would go undone, the world deprived of Emile's insight.

But something else happened inside him. Emile wrote to me that he was filled with "an awareness of all that is beautiful in the world." We all die, he said, but most of us don't know how much time we have. He was determined to fill his remaining days with community and purpose. Fresh out of surgery to remove a tumor from his brain, Emile convened a group of researchers at his home and issued a challenge. "Our goal should be more dramatic than just doing good science," he urged. Like Emile, they could go to war-torn places, speak with suffering people, and put science to work for peace. "We can walk through darkness and spread light."

Emile died on September 30, 2020. Many mourned an inspiring father, scientist, and friend. I also mourned his worldview. Emile believed hope is like light guiding our paths. If that was true, the world seemed to be dimming as the COVID pandemic lumbered on. Like the last, lingering moments of dusk, it was growing harder to see anything in front of us.

That year, the fault line separating my rosy persona and gloomy inner life widened into a canyon. Schools, hospitals, and companies invited me to speak about my work and help them feel hope, but mine was gone. On Zoom from my living room, I celebrated human kindness to people around the world. As soon as the screen went black, I returned to doomscrolling.

But my job is to be curious about the human mind, and after a while, I began to examine my own cynicism. It's a seductive worldview, dark and simple. Too simple, really, to explain much of anything. Cynicism encouraged me to expect the worst from people, but what gave me the right? It told me the future would be awful, but how could anyone know that? What was cynicism doing to me? To all of us? As I soon learned, it wears away the psychological glue that binds us. Trust, the willingness to be vulnerable to others, is an expression of faith that they will do the right thing. It is how hope lives between people. By eroding trust, cynicism steals our present together and dampens the futures we can imagine.

I thought often about Emile. How had he retained astonishing positivity even as his life was cut short? Can the rest of us do the same during our own dark times? These questions led me on a scientific journey that changed my mind, and a personal journey that changed my life. Exploring decades of research, I discovered that cynicism is not just harmful, but often naive. Hope and trust, by contrast, are wiser than most people realize. They are also skills we can build through habits of mind and action. I wish I'd known those practices earlier, but I am grateful for them now and believe they are worth sharing.

This book is about why so many people feel the way I used to, and how anyone can learn to think more like Emile.

The conversations he and I had in those hotel bars were not new. For thousands of years, people have argued about whether humanity is selfish or generous, cruel or kind. But recently, our answers have changed.

My parents immigrated to the United States in 1972. That same year, a project called the General Social Survey (GSS) began taking the nation's pulse, regularly polling people from every walk of life on a range of issues.

The country in which my parents arrived was no bed of roses. The Vietnam War was drawing down, but protests raged on. Operatives from the Nixon administration broke into the Democratic National Committee headquarters, leading to the Watergate scandal. Racial tensions ran high.

Yet, compared to now, 1972 America was a trust utopia. That year, nearly 50 percent of Americans surveyed by the GSS agreed that "most people can be trusted." By 2018, only 33 percent felt that way. If trust was money, its plunge would match the stock market's fall during the Great Recession of 2008. But unlike the economy, the trust recession has seen no recovery. Nor is mistrust merely an American problem. An international survey conducted in 2022 found that in twenty-four of twenty-eight nations, most people said their tendency is to distrust others.

Humanity has lost faith in humanity, and lost even more in our institutions. Between the 1970s and 2022, the percentage of Americans who trusted the presidency fell from 52 to 23 percent, newspapers from 39 to 18 percent, Congress from 42 to 7 percent, and public schools from 58 to 28 percent. Maybe we're right to suspect politicians and cable pundits. But our collective cynicism has consequences. Trust is not money, but it is just as vital for health, prosperity, and democracy. A run on the social bank can quickly collapse all three.

When trust is depressed, cynicism rises. Right now, it looks like an early frontrunner for mood of the 2020s. And why shouldn't it be? Our culture is flush with predators, Ponzi schemes, and propaganda. It's reasonable to decide people are interested only in themselves. But study after study finds that cynical beliefs eat away at relationships, communities, economies, and society itself.

This hurts people at nearly every level scientists can measure. Dozens of studies* demonstrate that cynics suffer more depression, drink more

---

* This book refers to *lots* of work in the social sciences, all of which can be found in the notes. If you would like to learn more about the research that backs up the claims here, see appendix B: "Evaluating the Evidence."

heavily, earn less money, and even die younger than non-cynics. In the seventeenth century, the philosopher Thomas Hobbes became cynicism's intellectual spokesperson. His book *Leviathan* argues that people need government to rein them in because left to our own devices, human lives are "nasty, brutish, and short." Few lines better capture a cynical view of life, but ironically, Hobbes's words best describe cynics themselves.

When I describe "cynics," you might conjure up a certain type of person: the toxic, smirking misanthrope, oozing contempt. But they are not a fixed category, like New Zealanders or anesthesiologists. Cynicism is a spectrum. We *all* have cynical moments, or in my case, cynical years. The question is why so many of us end up here even if it hurts us.

One reason is that our culture glamorizes cynicism and hides its dangers, through the promotion of three big myths.

Myth #1: Cynicism is clever. What is the opposite of a cynic? That's easy: a rube, chump, or mark, whose naive optimism sets them up for betrayal. This stereotype reveals what most people believe: that cynics are smarter than non-cynics. Most people are wrong. In fact, cynics do *less* well at cognitive tests and have a harder time spotting liars than non-cynics. When we assume everyone is on the take, we don't bother to learn what people are really like. Gullible people might blindly trust others, but cynics blindly *mistrust* them.

Myth #2: Cynicism is safe. Every act of trust is a social gamble. When we place our money, secrets, or well-being in someone else's hands, they have power over us. Most people who trust will get burned at some point. Those moments lodge themselves inside us, making us less likely to take chances again. By never trusting, cynics never lose.

They also never win. Refusing to trust anyone is like playing poker by folding every hand before it begins. Cynicism protects us from predators but also shuts down opportunities for collaboration, love, and community, all of which require trust. And though we forever remember people who hurt us, it's harder to notice the friends we *could have* made if we'd been more open.

Myth #3: Cynicism is moral. Isn't hope a privilege? Not everyone can afford to assume the best about people, especially if they have been harmed by a cruel system. In a world full of injustice, it may seem heartless to tell victims they should look on the bright side. Maybe optimists "hopewash" problems while cynics shed light on them.

This idea is intuitive but backward. Cynicism *does* tune people in to what's wrong, but it also forecloses on the possibility of anything better. There's no way to change a broken system if it's a mirror that reflects our broken nature. Why, then, do anything? At my most cynical, I felt morally paralyzed. I stopped volunteering and protesting, wondering why my more active friends even bothered. Other cynics tend to follow suit, sitting out elections and social movements more often than non-cynics.

Cynicism is not a radical worldview. It's a tool of the status quo. This is useful to elites, and propagandists sow distrust to better control people. Corrupt politicians gain cover by convincing voters that *everyone* is corrupt. Media companies trade in judgment and outrage. Our cynicism is their product, and business is booming.

Our beliefs influence how we treat other people, which shapes how they act in return. Thoughts change the world, and cynicism is turning ours into a meaner, sadder, sicker place. All of this is deeply unpopular. Americans trust one another less than before, but 79 percent of us also think people trust too little. We loathe political rivals, but more than 80 percent of us also fear how divided we've become. Most of us want a society built on compassion and connection, but cynicism convinces us that things will get worse no matter what we do. So, we do nothing, and they worsen.

According to an ancient myth, hope arrived on earth as part of a curse. Prometheus stole fire from the gods, and Zeus avenged the theft with a "gift." He commanded Hephaestus to mold the first woman, Pandora, and presented her to Prometheus's brother. Pandora, in turn, was given a clay jar—which Zeus told her never to open. Curiosity got the better of her, she lifted the lid, and out flew all the world's ills: sickness and famine for our bodies, spite and envy for our minds, war for our cities. Realizing her mistake, Pandora slammed the jar shut, leaving only hope trapped inside.

But what was it doing there in the first place, alongside our miseries? Some people believe hope was the jar's only good, and trapping it further doomed us. Others think it fits in perfectly with the other curses. The philosopher Friedrich Nietzsche called hope "the most evil of evils because it prolongs man's torment." You might agree. Hope has been typecast as delusional and even toxic—causing people to ignore their problems and the world's.

Scientists think of hope differently. The psychologist Richard Lazarus wrote, "To hope is to believe that something positive, which does not presently apply to one's life, could still materialize." In other words, hope is a response to problems, not an evasion of them. If optimism tells us things *will* get better, hope tells us they *could*. Optimism is idealistic; hope is practical. It gives people a glimpse of a better world and pushes them to fight for it.

Any of us can practice hope. Emile did. He saw the same world most of us do, but instead of retreating into cynicism, he chose to work for peace, build community, and live his principles. To me and many who knew him, Emile's positivity seemed supernatural. Temperament, experience, will, or some alchemy of all three graced him with a mind and a heart many of us could learn from.

This book is my attempt to spread his lessons. With his wife Stephanie's support, I've spoken with Emile's family, childhood friends, coaches, teammates, and colleagues. I've traveled to places that mattered to him and pored over notes he never got to share with the world. Through dozens of tearful, grateful conversations, I gained a deeper understanding of who Emile was and how he got that way. Then, unexpectedly, I began to experience his presence. When I felt snarky or cynical—which was often—I began hearing his voice: first occasionally, then often; first quietly, then clearly.

Shortly after his diagnosis, Emile wrote to Stephanie, "As a neuroscientist, I learned that our brains don't really see the world, they just interpret it. So, losing my body is not really a loss after all! What I am to you is really a reflection of your own mind. I am, and always was, there, in you." While writing this book, I've had the strange and solemn experience of witnessing Emile come alive inside my mind from beyond this world. He has taught me more than I could have imagined.

Here, he will teach you, too. Emile pursued peace the way doctors pursue healing. If illnesses are aberrations in the body's function, Emile saw conflict and cruelty as diseases of social health. He and his colleagues diagnosed the triggers that inspire hatred, and then designed psychological treatments to reduce conflict and build compassion.

Hope for Cynics will take a similar approach to our loss of faith in one another. You'll soon be able to diagnose symptoms of cynicism in yourself and others, understand its causes, and realize how it contributes to countless social ills, from an epidemic of loneliness to the "Great Resignation" at workplaces around the world to the erosion of democracy itself.

Once we understand the illness, we can treat it. In this mission, Emile becomes less like the doctor and more like a miracle patient. If cynicism is a pathogen, he was unusually resistant to it. When someone avoids a widespread plague, we might test their genes or blood for clues about how to fight the disease. I've probed Emile's life for choices and experiences that helped him practice hope.

In doing so, I've learned that one powerful tool he used to fight cynicism was *skepticism*: a reluctance to believe claims without evidence. Cynicism and skepticism are often confused for each other, but they couldn't be more different. Cynicism is a lack of faith in people; skepticism is a lack of faith in our assumptions. Cynics imagine humanity is awful; skeptics gather information about who they can trust. They hold on to beliefs lightly and learn quickly. Emile was a *hopeful skeptic,* combining his love of humanity with a precise, curious mind.

This mindset presents us with an alternative to cynicism. As a culture, we are so focused on greed, hatred, and dishonesty that humanity has become criminally underrated. In study after study, most people fail to realize how generous, trustworthy, and open-minded others really are. The average person underestimates the average person.

If *you're* anything like the average person, this hides some good news: People are probably better than you think. By leaning into skepticism—paying close attention rather than jumping to conclusions—you might discover pleasant surprises everywhere. As research makes clear, hope is not a naive way of approaching the world. It is an accurate response to the best data available. This is a sort of hope even cynics can embrace, and a chance to escape the mental traps that have ensnared so many of us.

Here, you will learn about decades of science on cynicism, trust, and hope, including work from my own lab, and meet people using hope like an axe to break down doors. We'll meet a principal who turned around a "dangerous" middle school by empowering its students and a CEO who replaced his firm's cutthroat culture with cooperation. A QAnon follower will discover that family means more to her than conspiracies, and a recluse in Japan will find his voice through art. In their stories, we will witness how our minds can evolve to strengthen communities and reimagine the future.

Throughout the book, I'll also share strategies and habits to cultivate hopeful skepticism. If you want to go deeper, appendix A offers a practical guide. But if I'm going to ask you to fight cynicism, I should take my own advice. Recently, I've tried. Drawing from the science, I've rethought parenting, experimented with the media I consume, talked with more strangers, and tried to overcome my climate "doomerism." Much of this work has been painful or awkward. But in fits and starts, it has changed me. I've watched my relationships strengthen, trust build, and optimism grow.

Cynicism often boils down to a lack of good evidence. Being less cynical, then, is simply a matter of noticing more precisely. I hope this book will help you witness the good in others and work toward the world most of us want.

The cynical voice inside each of us claims that we already know everything about people. But humanity is far more beautiful and complex than a cynic imagines, the future far more mysterious than they know.

Cynicism is a dirty pair of glasses more of us put on each year. I intend to help you take them off. You might be astonished by what you find.

Section I

UNLEARNING CYNICISM

Chapter 1

Signs and Symptoms

Cynicism is a disease of social health, but before treating it, we must understand what it is and how it affects us. Any diagnosis is detective work. Symptoms are its clues, outward signs that point to something wrong inside the body: Aches, tingling hands, and dizziness might signal anemia. Move that pain into the chest, and the cause might be more frightening. The meaning of each sign changes with the context.

Psychologists use people's words and actions as clues about their minds. If you no longer experience pleasure from your favorite activities, you might be depressed. If you are the life of every party, you're probably extroverted. We can diagnose cynicism this way, but it's tricky work, because the meaning of this word has morphed over time. Winding back through history, we discover that cynicism's origins have little to do with its dejected, current form.

Hidden Hope: The Ancient Cynics

The world's most famous fictional detective wasn't even the best one in his family. According to Sherlock Holmes, his brother Mycroft was more talented. Mycroft's problem was that he had "no ambition and no energy," and a disdain for humanity. Instead of solving cases, he created a hangout for people who didn't like people. As Sherlock describes it, the Diogenes Club "contains the most unsociable and unclubbable men in town." Any attempt to chat up a fellow patron could get you expelled.

The club was named after Diogenes of Sinope, an ornery Greek born twenty-three centuries earlier. The son of a banker, Diogenes was accused of counterfeiting his town's currency, went into exile, and lived on the streets of Athens, begging for food and sleeping in a large ceramic jar. He was less a classic philosopher than a counterculture stuntman, waging a full-frontal assault on polite society. He urinated, defecated, and masturbated in public. He waved a lantern in strangers' faces, claiming he was looking for just one honest person.

Equal parts monk, hippie, and insult comic, Diogenes terrorized some and fascinated more, who called him *kynikos*, or "doglike." He loved the name, saying, "I fawn on those who give, yelp at those who refuse, and set my teeth in rascals." *Kynikos* became the root of *Cynicism*. I'll call this ancient, original version "big-C Cynicism" from now on.

Diogenes gained a cult following. He and his fellow big-C Cynics were ironic, rude, and allergic to bullshit. But underneath, they preached hope. Cynics believed people were naturally capable of a virtuous, meaningful life, but rules and hierarchies robbed us of these gifts, poisoning us with cravings for wealth and power. Diogenes wanted to save people from these traps. As one scholar of Cynicism puts it, Diogenes "viewed himself as a physician who must inflict pain in order to heal." He didn't harass strangers out of hatred, but because he wanted to free them—like a Zen master slapping his student to startle them out of thinking.

To fight social illness, big-C Cynics created a recipe for living with meaning. Its first ingredient was *autarkeia*, or self-sufficiency. Ignoring convention, money, and status, Cynics could live on their own terms. Beholden to no one, they could pursue their true values. The second was *kosmopolitês*, or cosmopolitanism. Cynics rejected identity politics, seeing themselves as neither better nor worse than others. Asked where he was from, Diogenes answered simply, "I am a citizen of the world." The third was *philanthropía*, or love of humanity. Cynics responded to suffering with what one expert calls a "missionary zeal" to help others. "Concern for the well-being of one's fellow man is basic to Cynicism in all its forms," he writes.

Old-school Cynicism was the opposite of what it seemed. Under chaos, there was order. Under anger, care. Diogenes didn't avoid people; he tried to help them live truly and deeply. He probably would have despised the Diogenes Club.

How did his ideas become so twisted? Big-C Cynics preferred street theater to stenography, and their performances outlived their written record. As one historian wrote, "Cynicism's inability to give an account of itself" diminished its "persuasive charm." By not minding their legacy, Big-Cs let others write it through the lens of their own place and time. Some philosophers saw Jesus as an updated Cynic, with love for everyone and contempt for power. One Renaissance author cast Diogenes as a drunk, his ceramic jar full of wine.

Writers composed copies of copies of the philosophy. Cynics came to be remembered as malcontents—which they were—but their hope for humanity was left behind. Modern, "small-c" cynicism keeps the original suspicion of social rules but has lost its imagination and its mission. Big-C Cynics believed people had great potential. To small-c cynics, the worst elements of society reflect who we really are. Big-C Cynics mocked rules to escape them. Today's cynics sneer at society, too, but their detachment is a white flag of surrender—because to them, nothing better is possible.

A (Mistaken) Theory of Everyone

Small-c cynicism is the only form most of us know today; I'll just call it "cynicism" from now on. It infects more of us each year. To diagnose yours, think about whether you generally agree with these statements:

1. *No one cares much what happens to you.*
2. *Most people dislike helping others.*
3. *Most people are honest chiefly through fear of getting caught.*

In the 1950s, the psychologists Walter Cook and Donald Medley devised a test to identify good teachers. They asked hundreds of educators whether they agreed with these three statements plus forty-seven others. The more a teacher agreed, the worse their rapport with students. But the test had broader applications. The more statements *anyone* agreed with, the more suspicious they were of friends, strangers, and family. Soon it was clear that Cook and Medley had accidentally built an all-purpose cynicism detector.

Most people agree with between one-third and one-half of Cook and Medley's fifty prompts. I've simplified that into the few you answered above. If you disagree with all three, you're probably low in cynicism. If you agree with just one, you're on the low-medium end—think medium-rare for a steak. If you agree with two, you're on the medium-high end. And if you agree with all three, you might be a well-done cynic, with a bleak "theory of everyone."

We all use theories to explain, predict, and move through the world. Gravity is the theory that objects with mass attract one another. Even if you don't consciously think about it, this idea lives in your mind. It's why you're not confused when apples fall from trees, and why you probably think dropping a brick off a high-rise is illegal, but dropping a marshmallow might not be. Pretty much everyone shares a theory of gravity, but other concepts divide us. Optimism is a theory that the future will turn out well; pessimism is a theory it won't. Optimists pay attention to good omens and take risks; pessimists focus on bad signs and play it safe.

Cynicism is the *theory that people are selfish, greedy, and dishonest*. Like any theory, it changes how we see reality and react to it—in this case, the social world. In one of many studies like it, people took Cook and Medley's test, and then watched one person talk about their problems while another listened. Individuals who disagreed with Cook and Medley's statements rated listeners as warm and attentive. Those who agreed with the statements found listeners aloof and callous instead.

Cynicism changes how we think, what we do, and what we don't do. To further diagnose yours, let's try a game. Pretend you are an "investor" who starts out with $10. A second player, the "trustee," is a stranger whom you'll never meet. You can send the trustee as much of your money as you want. Whatever you pass along will be tripled. The trustee can pay you back as much of the money as they want. If you invest $10, it will become $30 in the trustee's hands; if they send back half, you'll both profit, each ending up with $15. They could also choose to send you all $30 or keep it all themselves.

Based on your first impulse, how much would you send? Write down your answer if you can—we'll come back to it in a moment.

Economists have used this game for decades to measure *trust*: one person's decision to put their faith in someone else. Every time you tell someone a secret or leave your kids with a babysitter, you make yourself vulnerable. If the people we trust honor their commitments, everyone wins. You confide in a friend, they listen and support you, and your relationship deepens. Your kids have fun with a new adult, the babysitter gets paid, and you enjoy sorely needed grown-up time. But people can also dupe us. Your new confidant might spread what you told him far and wide. The babysitter might steal from you or ignore the kids in favor of her phone.

Trust is a social gamble, and cynics think it's for suckers. Let's return to the game you just played. If you're like the average person, you would send about $5 to the trustee, which would then become $15. The average trustee would send you about $6 back, leaving you with $11 and them with $9 at the end of the game. If you're like the average cynic, you'd invest less, usually between $0 and $3. These choices reveal the theories we live by. Non-cynics think there's about a 50 percent chance the trustee will pay them back. Cynics think trustees will take the money and run. As it turns out, trustees pay back about 80 percent of the time. Cynics earn less than non-cynics in trust games, but almost all investors could earn more by trusting more.

In the lab, suspicion costs people money. In life, it deprives us of a much more vital resource: each other. The novelist Kurt Vonnegut wrote that people are "chemically engineered" to live in community, "just as fish are chemically engineered to live in clean water." Cynics, not wanting to lose, deny their social needs. They seek support from friends less often and negotiate as if the other party is trying to cheat. Like a trout washed ashore, they find themselves starving for connection.

This social malnutrition adds up over time. Studies find that cynical adolescents are more likely than non-cynics to become depressed college students, and cynical college students are more likely to drink heavily and divorce by middle age. Non-cynics earn steadily more money over their careers, but cynics financially flatline. Cynics are more likely to suffer heartbreak—and heart disease. In one study, about two thousand men filled out Cook and Medley's survey. Nine years later, 177 had died, and cynics were more than twice as likely as non-cynics to be among the departed.

In an old joke, two elderly women complain about the resort they're visiting. "The food at this place is terrible," says the first. "Absolutely," replies her friend, "and the portions are so small!" That might describe a cynical life: full of alienation and misery, and over too quickly.

Stalling Society's Engine

Cynics live harder lives than non-cynics, but as more people give up on each other, *everyone* pays the price. To understand how, we can compare the well-being of high- and low-trust nations. In 2014, the World Values Survey asked people around the globe if they agreed that "most people can be trusted." Fifty percent of Vietnam's citizens agreed, but in Moldova, which had a similar level of wealth at the time, a mere 18 percent did. Trust gaps occurred in richer countries as well, for instance, between Finland—58 percent trust; and France—19 percent.

High-trust communities lap their low-trust peers on many fronts. Their people are happier—in terms of well-being, living in a high-trust group is worth as much as a 40 percent pay raise. They are physically healthier and more tolerant of difference. They donate more to charity, are more civically engaged, and are less likely to die by suicide. They trade efficiently and invest in one another, allowing commerce to thrive. Economists once measured trust levels in forty-one nations, as well as their gross domestic product (GDP) over the following years. High-trust nations grew their wealth; low-trust countries' wealth stagnated or declined.

Trust makes good times better and bad times better, too. People who have faith in one another band together in the face of adversity. One dramatic example of this occurred in the Japanese city of Kobe. Two Kobe neighborhoods—Mano and Mikura—seemed alike on paper: barely three miles apart, both dense with factories, workshops, and houses; both home to aging middle- and working-class populations. But underneath the surface, these similarities disappeared. Mano was full of small family businesses, relying on trade between neighbors. Women played an integral role in its economy, whereas Mikura was more patriarchal.

The people of Mano had also faced challenges together. In the 1960s, a rising number of factories poisoned the air until 40 percent of neighborhood residents suffered from asthma. Public services such as garbage collection lapsed, and the streets were overwhelmed by rats, flies, and mosquitoes. Mano was slapped with an unwelcome nickname: "the department store of pollution." The population fell, and it looked like the neighborhood would become a slum.

Instead, residents fought back. They created a local planning committee and pressured the government for more resources and antipollution efforts. Slowly, parks appeared between the crowded streets. Factories relocated. Trash was collected. Soon children had places to play, and neighbors built homes for the elderly. The quality of life in Mano improved.

This activism bonded Mano's residents in common cause. Mikura, by contrast, lacked this history, and the trusting connections that came with it. Then, in 1995, a massive earthquake rocked all of Kobe and its surrounding areas. The tremors set off fires that lasted two days, claiming over five thousand lives and destroying more than one hundred thousand buildings.

As the flames spread, the differences between neighborhoods meant everything. Mikurans watched, many in their nightclothes, while homes turned to ash. Mano residents didn't wait for the authorities, but leaped into action together, forming pop-up bucket brigades, grabbing hoses from factories, and pumping water from rivers to fight the fire. About one in every four Mano homes were ruined—a terrible loss—but almost three out of every four were destroyed in Mikura. Mikura's death rate was ten times higher than Mano's.

During the quake, trust preserved buildings and the lives within them. In the tragedy's aftermath, it sped up recovery. Mano formed relief organizations, collected signatures to build temporary housing, and set up a makeshift day care. Mikura didn't coordinate and lost out on public services. The city of Kobe offered free debris removal if homeowners asked for it, but Mikurans didn't bother.

The effect of trust isn't confined to those two neighborhoods or that one disaster. Around the world, connections between people predict how well towns and cities bounce back from tsunamis, storms, and attacks. Networks of faith, community, and solidarity pivot in times of need, remaining agile and hardy. When communities lose trust, they grow unstable, like a Jenga tower with a bottom block knocked out. Crime, polarization, and disease rise.

The COVID pandemic put this in full view. In 2020 people's faith in government fell in the US and many other countries—but not everywhere. As the plague spread, South Korea's government sprung into action, following three principles: transparency, democracy, and openness. They invested heavily in rapid testing, regularly updating people on what officials knew (and didn't know) about the disease. This allowed them to quickly identify, trace, and provide government-subsidized treatment to sick individuals. South Korea's pandemic response earned citizens' trust, which they repaid in dividends. Infected people generally quarantined voluntarily without lockdowns. By the end of 2021, more than 80 percent of eligible South Koreans had been vaccinated, compared to barely 60 percent in the US and less than 70 percent in the UK.

As Prime Minister Chung Sye-kyun later reflected, "Once you have the trust of the people, it is possible to have a high rate of vaccination." The opposite was also true. Research found that distrustful people around the globe were less likely to be vaccinated, leading to more infection and death among low-trust nations and countries. According to one analysis, if every country in the world had experienced South Korea's high level of trust, 40 percent of global infection could have been prevented. But most countries were less like Mano and more like Mikura. The pandemic worsened cynicism, and cynicism worsened the pandemic.

Reviving Big-C Cynicism

If you came to this book for hope, you might think we're headed in the wrong direction—confirming your sense that the world is getting worse. But what goes down can come up. As we will witness many times, trust can and has been rebuilt. Ironically, some treatments for modern cynicism emerge from its big-C roots. Diogenes's principles—self-sufficiency, cosmopolitanism, and love of humanity—can be a starting point for cultivating hope. My friend Emile is a striking example of how that can work.

On the surface, Emile was Diogenes's photonegative: warm and tolerant where the Greek was prickly and sour, a coach and teammate rather than a loner. Yet the two had a lot in common. Diogenes rejected wealth; Emile never had it in the first place. Both lived with unusual amounts of freedom. In Emile's case, this came from his father, Bill, an author, gardener, bookstore clerk, and consummate dabbler. As a young man, Bill had bounced around the Bay Area, as he puts it, "on the margins of society—until I became a father. That changed everything."

With Emile's mother too sick to parent, Bill raised the boy alone. He would plop baby Bruneau into a refrigerator box full of stuffed animals from Goodwill and tow him by bike to sidewalk cafés and through local forests. As his son grew, Bill was a constant presence, but rarely told him what to do. Emile later called this parenting style "underbearing attentiveness." "The remarkable gift my father gave me," he wrote, "was to allow me to grow into myself—to become me."

Emile developed an abiding disinterest in money and status, even though the Bay Area towns around him were crammed with both. "He had nothing to lose," a close friend remembers, "because he was happy with nothing." This freed him—Diogenes-style—to roam through life on his own terms, following whatever called him. At Stanford, he played on the men's rugby team, and in his free time would sit for hours with unhoused locals, an unusual habit in Palo Alto's gentrified neighborhoods.

After graduation, he taught science at a wealthy prep school, but soon got fed up with its glitzy fundraisers. He left, moving to Michigan to pursue a PhD in neuroscience. In hopes of understanding his mother's illness, he spent years examining slices of brain tissue from deceased patients who had lived with schizophrenia.

In his spare time, Emile traveled voraciously. One summer, he spent weeks at a camp designed to promote peace between Catholic and Protestant teenagers in Ireland. The boys wiled the summer away together, playing and sharing bunks and meals. But on the very last day of camp, a fight broke out. The kids immediately fell back into their religious tribes, undoing the camp's efforts in an instant. As a counselor separated two brawling boys, one of them screamed to the other, "You orange bastard!," referring to William of Orange, the seventeenth-century king of England. The stain of past wars lived inside these children, and a friendly summer wasn't going to help, any more than a Band-Aid applied to a third-degree burn.

This was a pivotal moment in Emile's life. The camp's failure made him despondent, then resolute. He had seen how schizophrenia disrupted the brain, and was set to join hundreds of scientists trying to help people like his mother. Now he realized that hatred was a brain disease, too, one that warps people's minds and drives them to stunning cruelty. But unlike schizophrenia, hate wasn't a blockbuster topic in brain research. And without understanding it, how could he help overcome it?

Emile committed to studying the neuroscience of peace. There was just one problem: That science didn't exist. So, he convinced a renowned researcher at MIT to help him build it. Emile and his new mentor used MRI scanners to probe what happened in the brains of Palestinians and Israelis as they read about one another's misfortune. His work brought him to Europe to study the Roma, to Chicago to meet with former white supremacists, and to Colombia to treat the scars of civil war.

Emile's interests didn't conform to one clear category, and he showed little interest in staying within other people's borders, either. As a child, he despised shoes, going mostly barefoot until seventh grade, when his new school required footwear. Not owning any, he borrowed his stepmom's sneakers. Emile was rarely hurried and enjoyed getting lost, even when his travel partners had somewhere to go. As one of his mentors told me, "He wasn't a person you could 'manage.'"

Emile also refused to compromise his values for the sake of convention, whether his choices were large or minuscule. Every time he and Stephanie went out for dinner, Emile brought Tupperware for leftovers to avoid single-use plastic. "Sometimes it was exasperating, but it was always admirable," she remembers. "Emile had a very strong internal compass and a commitment to that compass."

Trusting Ourselves; Listening to Others

Emile lived the big-C principle of *autarkeia*, or self-sufficiency. I don't know if he was a Diogenes fan, but he loved another thinker who spun *autarkeia*'s modern remix. One of the few possessions Emile cherished was a hand-scribed copy of *Self-Reliance* by Ralph Waldo Emerson, which he kept in a glass-fronted box on his bedside table.

Emerson didn't urinate in the town square, but he loathed convention as much as any big-C Cynic. "Society everywhere is in conspiracy against the manhood of every one of its members…" he wrote. "It loves not reality and creators, but names and customs." Like Diogenes, Emerson thought the way out of this trap was to follow our hearts without compromise or fear. "In self-trust all the virtues are comprehended," he wrote.

On the book review website Goodreads, Emile said this about Emerson's work:

> The essay "Self-reliance" remains one of the most influential pieces of material that I have had to guide me in the development of my own character… It gave me a call to arms and inspiration to become a good and true man while trusting me to determine who that man will be.

This review threw me for a loop. I'd always seen Emile as intensely oriented toward other people, something I thought we shared. And he *was* that way. Several people I spoke with brought up how he listened, so intently that you felt yourself come into focus through his eyes. His social media posts, even about contentious political issues, brim with humility.

How did this square with fierce self-reliance—even a belief that society is a conspiracy against its members? To me, togetherness is the best of our species. The worst often comes when people trust their internal compass too much. Conspiracists, racists, and demagogues don't care what you think about them. Their confidence drowns out everyone else. Wouldn't we be better off if they doubted themselves more?

I spent several nights troubled by this, and then realized the answer—like Emile's childhood—was just a few miles away, in the research my Stanford colleague Geoff Cohen does on beliefs and values.

You might think that beliefs and values are like chocolate and dark chocolate—different flavors of the same thing. In fact, they're quite different. Beliefs are assumptions or conclusions; values are the parts of life that bring a person meaning. Beliefs reflect what you think of the world; values reveal more about yourself. Confusing these two can be dangerous business. When someone attaches their self-worth to a belief—political, personal, or otherwise—they desperately need to be right. Challenges to what they think feel like threats to *how* they think—evidence they aren't smart or good enough. The person screaming loudest is often most fearful of being wrong.

Though cynics doubt others, they also tend to define themselves through social comparison. In one study, people who agreed with Cook and Medley's bleak statements about humanity were also more likely to say they depended on prestige and status for a sense of self-worth, and to worry they didn't measure up socially. Needing to prop themselves up, they searched for evidence that could put others down.

One way out of this trap is through focusing on our deepest values, very much like *autarkeia*. In Geoff's studies, people are shown a list of qualities—for instance, social skills, close relationships, and creativity. They are then asked which one matters most to them and told to "affirm" this value in their own lives. If you ranked being funny as important, you might then write a paragraph about "personal experiences in which your sense of humor was important to you and made you feel good about yourself."

When people affirm what matters most to them, they are reminded of their highest purpose, which makes everyday social threats less dire. Geoff's studies and many others find that people who affirm their values become *more* open to information that contradicts their beliefs. It takes confidence in yourself to question your opinions. Among adolescents, values-affirmation also increases kindness toward others and trust in schools. By connecting us to ourselves, affirmations calm cynicism.

Perhaps because of his dad, it seemed natural for Emile to articulate and express his values. But for many of us, doubt starts at home. A person who doesn't have a strong hold on their values can feel internally flimsy, grabbing on to baubles like praise and prestige to steady themselves. "Yes, we are the cowed—we the trustless," Emerson wrote.

I recognize this all too well. For as long as I can remember, I've worried about my place among others. I'm hopeless at team sports but also calculus. As I discovered other strengths and, to my surprise, gathered some successes, it was easy to stack these up as stand-ins for self-worth. This put me in a reliable state of threat. The more I counted on looking smart, the more I feared being dumb. When someone challenged my scientific ideas, I could have engaged, but often got defensive instead. When someone else published a wonderful new experiment, I should have felt interest and delight, but again and again felt jealousy rising in their place.

That changed when my daughters were born. My care for them overwhelmed any need to prop myself up. Becoming a dad was the spiritual equivalent of wearing contact lenses for the first time: The world sharpened with details I never knew I'd missed. A surge of love made the posturing and politics of life as a professor seem small and ridiculous. Instead, the wonderful colleagues and breathtaking ideas that had been around me the whole time came into view. The kids were bundles of curiosity. Watching them, my own grew as well.

Parenting straightened my inner compass. For others, true north appears through the purpose of a dream job, the thrill of a new romance, or the clarifying sadness of loss. But it doesn't take earth-shattering events to tap into our values. Geoff's work shows us that through simple exercises, we can get closer to them anytime we want. As Diogenes, Emerson, and Emile remind us, if we want to rebuild trust in our relationships and communities, we must also trust ourselves, listening to the voice that speaks to us when the rest of the world is silent.

Chapter 2

The Surprising Wisdom of Hope

If cynicism were a pill, its warning label would list depression, heart disease, and isolation. In other words, it'd be a poison. So why do so many of us swallow it? One reason is that many people think cynicism comes with another, more positive side effect: intelligence.

Imagine two individuals: Andy and Ben. Andy believes that most people would lie, cheat, or steal if they could gain from it. When someone acts kindly, he suspects ulterior motives. Ben thinks most people are altruistic and would *not* lie, cheat, or steal. He believes people act selflessly out of the kindness of their hearts.

Knowing only what you've read so far, whom would you pick for each of these assignments: Ben or Andy?

1. Write a powerful argumentative essay.
2. Take care of a stray cat.
3. Calculate interest on a loan.
4. Cheer up a lovesick teenager.

If you picked our cynic, Andy, for tasks 1 and 3, and Ben for 2 and 4, you're like most people. The odd-numbered jobs here are *cognitive*, requiring precise thinking; the even ones are *social*, requiring the ability to connect. Researchers recently asked five hundred people to choose a cynic or a non-cynic for many tasks like these. More than 90 percent chose Ben for social tasks, but about 70 percent chose Andy for cognitive ones. They acted as though non-cynics are kind but dull, and cynics are prickly but sharp.

Most people also think cynics are *socially* smart, able to slice through insincerity and dig out the truth. In one study, people read about a company where new employees had lied to get their jobs. Readers were asked to assign one of two managers, Sue or Colleen, to handle interviews moving forward. Both were equally competent, but Sue "view[s] people very positively, and her default expectation is that everyone she meets is basically trustworthy." Colleen begs to differ; she thinks "people will try to get away with everything they can." Eighty-five percent chose Colleen as the new interviewer, confident she'd be better at spotting liars.

More than a century ago, the writer George Bernard Shaw quipped that "the power of accurate observation is commonly called cynicism by those who haven't got it." People who count on Andys and Colleens agree. A sucker is born every minute, but if you circle life's block enough times, you learn not to trust everyone, and eventually to trust no one.

Over the last few years, I've met dozens of self-proclaimed cynics. Besides the obvious contempt for people, most have something else in common: a harsh pride. It may feel better to believe in people than to be cynical, they say. But we can't go around thinking whatever we want, just like we can't pretend tiramisu is a health food. Cynics might live hard lives, but that's just the price of being right.

If cynicism is a sign of intelligence, then someone who wants to appear smart might put it on, like wearing a suit to a job interview. And indeed, when researchers ask people to appear as competent as possible, they respond by picking fights, criticizing people, and removing friendly language from emails—performing the gloomiest version of themselves to impress others.

Most of us valorize people who don't like people. But it turns out cynicism is not a sign of wisdom, and more often it's the opposite. In studies of over two hundred thousand individuals across thirty nations, cynics scored *less* well on tasks that measure cognitive ability, problem-solving, and mathematical skill. Cynics aren't socially sharp, either, performing worse than non-cynics at identifying liars. This means 85 percent of us are also terrible at picking lie *detectors*. We choose Colleens to get to the bottom of things when we should join team Sue.

In other words, cynicism *looks* smart, but isn't. Yet the stereotype of the happy, gullible simpleton and the wise, bitter misanthrope lives on, stubborn enough that scientists have named it "the cynical genius illusion."

Skepticism: The Scientific Mindset

Cynics often get people wrong, but that doesn't mean it's smart to put faith in everyone, all the time. Researchers once measured trust in hundreds of children, and then checked in with them a year later. When it came to depression and friendships, cynical children ended up the worst off, but extremely trusting kids did less well than those in the middle.

Why? When judging humanity, both cynics and trusters behave like lawyers in humanity's trial. Trusters represent the defense. They disregard suspicious signs, forget betrayals, and hold on to any evidence of human goodness. Cynics work for the prosecution, explaining away kindness and cataloging every instance of human viciousness. Both attorneys are inclined to ignore half of the evidence, even if they're opposite halves.

Lawyering is a fine way to argue, but a terrible way to learn. A growing science finds that real wisdom arrives when people know what they *don't* know. Likewise, social wisdom doesn't mean believing in everyone or no one. It means believing in evidence—by thinking less like a lawyer and more like a scientist. And despite the different instruments they use, all sorts of scientists share an intellectual tool: *skepticism*, the questioning of old wisdom and hunger for more information. Skeptics update their beliefs based on new information, allowing them to adjust to a complex world.

Researchers recently asked hundreds of people about their cynicism and skepticism (for instance, "Before I accept someone's conclusion, I think about the evidence"). They found that someone's level of cynicism didn't necessarily predict how skeptical they were, or vice versa. And whereas cynics were more likely to fall for conspiracy theories, skeptics were *less* prone to these mental stumbles.

Instead of thinking of social wisdom along one dimension—is someone cynical or trusting?—let's consider two dimensions: someone's faith in people, and their faith in data. This creates four general ways people could respond to others:

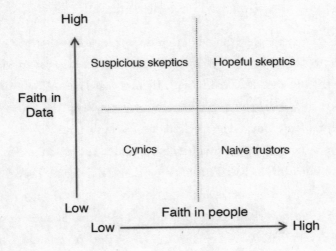

At the bottom of this graph are the lawyers in humanity's trials. On the bottom left are cynics, who are sure people are awful. On the bottom right are naive trusters, just as sure about others' good intentions.

As we move from the bottom to the top of this space, a person thinks more like a scientist, using skepticism to decide what they believe. On the left are suspicious skeptics, who start out negative but stay open-minded. That's where I tend to land. I fear the worst about people, but as a scientist

I am also restless in my assumptions. On the top right are hopeful skeptics, like Emile. He was perpetually curious, but with a more positive default than mine.

Where are you in this space? A person could start out anywhere. One hopeful skeptic might be more positive (farther to the right on this chart) than another, and one cynic might be more naive (farther down) than another. We can also move freely through the space. Skeptics are especially adaptable. Put a hopeful one in a high-stakes poker game, and they will start looking for bluffs. Move a suspicious one to a friendly block and they'll start counting on their neighbors. Less-skeptical people have a harder time adapting, but that doesn't mean they can't.

The cynical genius illusion reveals just how harmful this worldview can be: Cynics end up sicker, sadder, poorer, *and* more wrong. So, if cynicism isn't a matter of cleverness, why do people turn toward it? Under all that bluster, many cynics are just trying to stay safe in the face of suffering.

Disappointed Idealists

By eight o'clock that morning in June 2020, Megan was not who she'd been the day before. She had spent all night watching videos inspired by the QAnon conspiracy theory and believed the world's problems were caused by a cabal of twelve oligarchs, descended from the same Sumerian bloodline, who ran a global child trafficking network. Her new world was terrifying, but she felt joyous. "The grace of God was upon me," she recalls.

Megan shared her insights on Facebook, posting that Bill Gates was taking advantage of the COVID crisis to surveil humanity. "He's using microneedling tattooing vaccines so that a person can be scanned the same way groceries are scanned," she warned, providing a string of YouTube videos as "proof." People attacked, mocked, and unfriended her. Megan's boyfriend worried out loud that she might kill him in his sleep. The reactions distressed her, but online she remained eerily diplomatic. "First of all," she replied to one person, "while my perspective is different...it in NO WAY changes the love, respect, and appreciation I have for you."

Megan could afford to be polite because she was confident, referring to herself as "red-pilled." The term is drawn from *The Matrix*, in which Keanu Reeves's Neo chooses between a blue pill that will continue his blissful ignorance and a red one that will reveal the dark truth. Red-pillers come in many forms, from conspiracy theorists to hyper online misogynists, but they share a common belief: The more you know, the worse people look. Megan was sure most of the world was asleep. Her new mission was "to help them wake up."

Megan was an unlikely candidate for the QAnon movement. A self-proclaimed California Progressive, she studied healing meditation and nonviolent communication, a technique to resolve disagreements through empathy. During conflicts, she searched for "unmet needs"—like being heard and cared for—hiding under people's outbursts.

Megan was attuned to emotional hunger because she had felt it for most of her life. Growing up in the Midwest, she drew warmth from a deep connection to her father, Harold. Harold was an eternal optimist who "flowed with whatever life brings," and showered Megan with affection and praise. Her mother, Eileen, was intense and distant.

Shortly after Megan's ninth birthday, Eileen kicked Harold out. He didn't want to leave, but "flowed" with this, too. Megan watched numbly as his little green Honda turned left and then disappeared down the road leading away from their home. Within a year, he had a new girlfriend and lived two thousand miles away. Harold had provided the joy and safety in Megan's childhood. To her, the divorce was like an amputation. As soon as she could, she moved to California to be with Harold, ricocheting across the country to live with each parent for stretches of high school, unable to truly count on either of them.

Megan saw Harold as the victim and Eileen as the authority figure who had broken their family. Rebellion against her mother quickly bloomed into suspicion of teachers, doctors, and anyone in power. Some of Megan's beliefs were pretty standard (politicians are "bought and paid for," she tells me), some less so (9/11 was an "inside job"). In 2016, she experienced a rare glimmer of hope, in the form of the Bernie Sanders presidential campaign. Megan loved the candidate's authenticity and shared his horror at wealth inequality. That summer, she handed out postcards and T-shirts decorated with Bernie's toothy grin and posted "five times a day" on Facebook about how much the country needed him.

When Sanders lost the primary in 2016, and again in 2020, Megan's last bits of faith in the system imploded. Then the pandemic hit. A natural extrovert, she wilted in lockdown. Her boyfriend, Thomas, tumbled into depression following the police murder of George Floyd. Their apartment—like so many that year—became stifling and claustrophobic. Amid a string of bleak days, a close friend texted Megan asking her to watch a video called *The Fall of the Cabal*. The timing couldn't have been better, or worse.

Some conspiracy theorists are drawn in by bigotry and violence. Not Megan. She knew, in her bones, that something was wrong—in her life, and in the world. Q gave a name to that fear. The world was broken, but at least she knew how, and was sure a group of heroes would soon redeem it. More than anything, Q made her feel less alone. Old friends rejected her dark fantasies, but fellow believers quickly took their place, commending her for peeking behind the curtain with them. "It was a breath of fresh air," she said. "I got eye contact, deep listening, and empathy. It was such a contrast to what I was getting anywhere else." The Q community was a Harold-like oasis in a world of judgmental Eileens.

The psychologist Karen Douglas has studied conspiracy theories for over a decade. She writes that many people gravitate toward this type of thinking "when their existential needs are threatened, as a way to compensate for those threat[s]." Compared to nonbelievers, conspiracy theorists tend to be more anxious, feel less control over their lives, and report weaker connections to family. Like individuals who have lost the hold on their values, they grip more tightly to their beliefs, no matter how unrealistic.

The place where Megan fulfilled her needs happened to be a cultlike community that stokes violence and contributed to the January 6 insurrection. Conspiracy theorists do vast amounts of harm to themselves, their families, and society. Understanding their reasons for joining doesn't excuse anyone. But a bit of curiosity complicates the story most of us tell about fringe ideologies, and sheds light on how cynicism eats away at people.

I don't know what Megan was like as a one-year-old, but by that age most people have decided whether they can trust the world.

In the 1970s, psychologist Mary Ainsworth brought mothers and infants to a laboratory playroom. A stranger joined them, then the mother left her child alone with the stranger for a minute or two before returning. Maternal abandonment, however brief, is a stressful surprise for any baby, but Ainsworth found children reacted in different ways. About two-thirds rolled with the punches. These babies happily explored a new space with their mom, freaked out when she left, and soothed quickly when she returned. Ainsworth referred to these children as "securely attached." In contrast, the other third were "insecurely attached": nervous even in their mother's company, inconsolable when she left, and still upset after she returned.

None of these children remembered their experience in the lab, because no one remembers being one year old. But as Ainsworth found, our first year on earth leaves imprints deeper than memory. Securely attached kids learn they can trust their caregivers. By extension, the world becomes safe and full of possibility. Insecure children learn the opposite, and that instability reverberates through their lives. As adults, insecurely attached people are more likely to distrust lovers, friends, strangers, and institutions. What's more, insecure attachment has spread since Ainsworth studied it, increasing by about 8 percent among Americans between 1988 and 2011. It's impossible to tell if this trend has worsened the US trust deficit, but it probably hasn't helped.

Still, the story is complex. Insecure attachment comes in different forms: Some people hold on tightly to loved ones, terrified to lose them. Others act aloof, sure people will leave them no matter what they do. A person can be secure in some of their relationships but not others: counting on family but doubting romantic partners or vice versa.

I can't imagine joining QAnon like Megan did, but the early disconnection she described resonated with me immediately. My parents come from Peru and Pakistan, two nations separated by ten thousand miles and just as many cultural differences. Somehow, my mom and dad have even less in common than their home countries. When I was eight years old, they informed me they were splitting up. I didn't ask why, but wondered why they'd gotten together in the first place. Much of my childhood dissolved in the acid bath of their long divorce. I barely remember anything before my twelfth birthday, and when I do most of what comes back is a parade of stony silences, bitter outbursts, and night after night alone.

My parents each did their best, but both were caught up in a mess of emotions, which crashed through their homes like tumultuous weather fronts. I treated being a son like walking on a balance beam—tiptoeing across each day vigilant for what each adult wanted from me, terrified of screwing up.

An aching need to be worthy of my parents spread to other relationships. In school, I clung to close friends. When they spent time with someone new, I fretted about being left behind. Later, relationships with women eclipsed every other kind. I poured myself into romance with a jittery eagerness, often before getting to know a new girlfriend well. Occasionally this devotion charmed someone; more often it drove people away. I projected a plucky, curious persona to be interesting and attractive, but it rested unsteadily on layers of nerves. A philosophy clawed at me from childhood: Life is a competition for people's love. No one cares unless you make them.

My insecurity was different from Megan's. Where she felt hostile toward the world, I was fearful. But we both transformed early pain into our own theories of everyone, varying flavors of cynicism that stuck to us decades after we left home.

The comedian George Carlin once said, "Scratch a cynic and you'll find a disappointed idealist." Megan and I—and countless others—turned to cynicism to recover from hurt. But suffering can be *too* good a teacher. A puppy that's been abused will remain skittish around new people, even if they mean him no harm. One toxic romance or teenage bully can destroy a person's trust for years.

After traumas and betrayals, a skeptic could rightly lose faith in the person who hurt him. In the two-dimensional space we charted above, they would move to the left, becoming more guarded but remaining open to new people. Instead, victims often become *pre-disappointed*: generalizing from bad experiences and deciding that no one can be trusted, moving down and to the left on our chart, toward cynicism.

Let's revise Carlin's law: Scratch a cynic and you'll find a *pre*-disappointed idealist. Reeling from pain, they give up on curiosity to defend themselves. This is an understandable response to pain, but it also prevents cynics from turning strangers into friends, confidants, and soulmates. Cynics land in a negative feedback loop. Their assumptions limit their opportunities, which darken their assumptions even further.

How can they get unstuck?

A Safe Home Base

Parenthood changes life forever. For me, it created space to let go of others' approval. For Emile's mother, Linda, change came in a harrowing, uninvited way. Shortly after giving birth, Linda was plagued by cruel, demonic voices that mocked and accused her—the torment of schizophrenia. Trapped in her own mind, she left Emile and Bill and lived on the streets of Palo Alto. Unsheltered and alone as a twenty-five-year-old woman, she suffered chronic abuse. When she sought treatment, 1970s-era psychiatry was often abusive, too.

In a sad irony, the arrival of her beloved child biologically triggered the illness that prevented her from being with him. Linda would appear in Emile's life occasionally—sometimes confused and disheveled. In many ways, their early relationship was utterly insecure. This could have lodged itself inside Emile. He could have felt ashamed of Linda or decided her suffering was his fault. He could have concluded life is ugly and unfair, and come to resent people who had it easy. But none of this happened. Why?

The answer starts at home. Despite being poor and single, Emile's dad, Bill, was doggedly present with his son, offering the "underbearing attentiveness" Emile cherished. And though their time together was unpredictable, Linda and Emile clearly cared for each other. Outside his house, just before the two would meet, she would sometimes be visibly distraught, fighting the voices. Then, through force of will, she'd compose herself for as long as they were together. Family members and friends recall their reunions as peaceful and affectionate. Mother and son carved out a small space, away from a difficult world and the devils in her mind. There, they did what parents and children do best: love one another.

Linda died when Emile was in his thirties. He now lived across the country, and in her last year he flew to be with her, advocating for her with doctors during the day, sleeping on her hospital room floor at night—parenting her in a way she could never do for him. After her death, Linda lived on in his memory, not despite her pain but because of it. He lacked a "normal" mother but had found a hero, and the beginnings of his worldview.

Several people told me that Emile's relationship with Linda marked him with inner "superpowers." As one friend remembers, "He understood from early on that wonderful people could end up in terrible circumstances through no fault of their own." Curiosity came naturally to him; snap judgments felt foreign. This skepticism formed the bedrock of Emile's career as a scientist, and lent him patience in his work as a peacemaker.

Many people live in the ruins of their worst moments, trapped by pre-disappointment, cynicism, and loss. For Emile, suffering popped up an antenna, tuning him in to what others endure, building his compassion. The power of adversity to build kindness is old, and everywhere. In the weeks after an assault, most survivors report greater empathy. Members of war-torn communities cooperate more freely than those from other towns. In laboratory studies, people who have gone through greater suffering are more likely to help strangers.

Why does pain close some people off and open others up? Lots of things matter here, but one is community. In Mano, the high-trust Japanese neighborhood, an earthquake catalyzed cooperation. In the less-trusting area of Mikura, things fell apart. During times of adversity, lonely people often become lonelier, trauma curdling into pre-disappointment. When—like Emile—someone has support, they have a better chance of growing through hardship.

In 2019, with a surgical scar running along the base of his skull, Emile spoke to friends and colleagues about his own children, who would soon be without a father. "Trauma can have bad or amazing effects on children," he said, "and it all depends on the environment they're in." He and Stephanie focused on creating a home in which their kids could grow through pain. Each child had responded to the news of Emile's sickness in different ways. Their six-year-old daughter, Clara, crawled into a linen closet, climbed onto a shelf, and shut the door. Their four-year-old son, Atticus, began talking obsessively about death—in what order the family would die, and what he'd do when he was the last one alive.

Stephanie and Emile responded to all this like his dad, Bill, might have, with underbearing attentiveness. Emile gathered boards, a hammer, and nails, and together with Clara built a nook in the linen closet, with steps, a railing, a flashlight, and her favorite stuffed animals. "Instead of the shelf becoming a retreat from me," he wrote, "it became a retreat from this [the stress of his illness]." When Atticus brought up what he'd do when Emile died, Stephanie asked if she could join him, and they made a map together, of where they'd visit and how they'd honor him. The parents met each kid where they were. If trauma passes through generations, Emile and Stephanie were determined to pass on love and strength along with it.

Not all parents can offer such sublime devotion to their children, and not all children are blessed with underbearing attentiveness. But even for the rest of us, insecure attachment is not a life sentence. By working on ourselves and through the grace of new relationships, insecure people can achieve "earned attachment": a sense of safety and connection later in life. We can build trust and hope from the ground up.

For Megan, who had fallen prey to the QAnon conspiracy, change arrived through chosen family. After his initial shock, her partner, Thomas, decided he would stay with her. He never pretended to accept Q but was fierce in accepting her. "I don't believe what you believe," Thomas would tell Megan, "but I know your heart and I love you." Her father, Harold, also remained close to Megan and showed a similar openness.

Many people have tried just as hard as Thomas and Harold to connect with struggling family members and still lost them to conspiracy theories. But for Megan, a low point became a turning point. In the safety of her closest ties, she started to rebuild her security.

Skepticism About Cynicism

A safe home base changes how people feel, and how we think. Skepticism can seem like a cerebral exercise, and wisdom like a solitary virtue, but both live in the space *between* people. Like the children Mary Ainsworth studied, people can afford to be curious when they feel safe. In one of many studies like it, religious people read arguments for atheism, and vice versa. Securely, versus insecurely, attached individuals were more open to information that challenged their beliefs. In a twist, researchers then asked some people to think about times in their childhood when they felt secure and "didn't have to worry about being abandoned." People who recalled comforting moments became more open-minded than people who thought about tough times from their childhood, or nothing at all. Like the core values Geoff Cohen studies, deep personal connections give us the space to let go of rigid thinking.

Cynicism can spread between people, but so can skepticism. Even in the face of deep disagreement, a person who shows openness can help others feel safe, which in turn opens their minds as well. Thomas tried this with Megan. "I'm willing to consider the possibility I might be seeing things the wrong way," he told her. "Are you willing to do that as well?" He agreed to listen to an hour of her best Q evidence each week, challenging her to make the strongest possible case. Megan jumped at this deal, sure she would "awaken" Thomas. But as she pored over Q materials, something felt different. Before, she'd read them eagerly, knowing she'd be passing "insights" back and forth with fellow QAnons, like a red-pilled book club. Now, when she read she felt Thomas—and his skepticism—next to her. "That gave me motivation to look more deeply," she remembers, "rather than just being influenced by clickbait."

Megan became skeptical *of* her cynicism. She had always doubted politicians; why not treat QAnon the same way? When she did, conspiracy logic quickly crumbled. Q promised that politicians and Hollywood stars would soon be arrested, but the day never arrived. Conversation around child trafficking quickly switched to election fraud, a Chinese Communist takeover of the US, and coming food shortages, as though the community were picking news stories out of a hat. "There were so many things I was falling for," she remembers, "and it didn't take much to put holes in them. I just needed motivation to look for those holes." Finally, one day in December, she couldn't believe it at all anymore. She quietly left QAnon and began repairing her life.

My own youthful cynicism sprang a leak during graduate school in New York City. A new girlfriend (now my wife of twelve years) upended my assumptions about relationships. Landon was notably uninterested in the performance I dangled in front of everyone, and notably present when I was less entertaining. Weeks into our relationship, my grandmother died. I was visiting Washington, DC, when the news came. My family lives in Boston, so reaching them meant taking buses up the Northeast Corridor. Landon met me halfway along the trip at New York's Penn Station. We sat at the bar of an all-night diner and she asked me about my grandmother's life. I was a distraught, unattractive mess—leaking emotions I would have normally hidden. Her easy, solid warmth never faltered. In her presence, I felt a calm that had eluded me for most of my life.

This didn't flip my attachment switch right away. In the months that followed, insecurities popped out of me, threatening to drive Landon away. But I was determined that this time could be different. So, like half of Manhattan, I started therapy. My psychologist probed the anxious beliefs inside me like a physician pressing for organs under the skin. Why did I think people would leave the moment I was no longer interesting? She challenged me to defend my cynicism like I would defend a scientific hypothesis—with data. And if I didn't have data (I usually didn't), she encouraged me to gather some. What would happen if I let down my guard once, twice, or even most of the time?

My deepening connection to Landon gave me the safety to try those experiments, replacing cynicism with hopeful skepticism bit by bit. Strangely, this might have made me look *more* cynical to outsiders. In the past, I'd worked hard to please everyone with a rigid positivity. An acquaintance had nicknamed me "Guy Smiley"; this was not a compliment. Being more authentic often meant opening up about something I didn't like, someone I didn't trust, or the dark mood I was in. It made me more vulnerable, and this resulted, as far as I could tell, in exactly zero people abandoning me. The data were clear. Guy Smiley could retire. Slowly, with help, I reshaped my assumptions. According to a recent test, I remain insecurely attached to my parents, but am secure in my relationship to Landon, and in general.

Finding Social Wisdom

Skepticism and cynicism aren't just different. The first can be an antidote to the second.

For more than a century, skepticism has been used as a healing tool in cognitive behavioral therapy, or CBT. CBT providers team up with patients to challenge their assumptions, just like my therapist did. In *reality testing*, patients identify and express exactly what they believe. An anxious person might think that deep down, their friends hate them. The therapist and patient then fact-check those feelings. Has *anyone* ever said they like the patient? Been kind to them? Asked them to spend time together? In almost every case, real-life evidence contradicts black-and-white assumptions.

The next step is to act like a scientist, using CBT's *behavioral experiments*. Therapists and patients will agree on how the patient can test his beliefs. Someone who thinks everyone hates him might ask a few friends to go to the movies. If the anxiety is right, no one will show up. If things turn out another way, the patient can rethink their assumptions—like Megan and I did.

Depression and anxiety blanket us in negative assumptions about ourselves. Cynicism does the same for our beliefs about one another. Both are rooted in pain, and both freeze us in place. But underneath that bad news is good news. Cultivating a more positive outlook doesn't require anyone to force a smile, lie to themselves, or "fake it until they make it." If you're like most people, your starting point about other people is probably too negative. Thinking like a scientist, you might pick up some hope along the way.

I am still working on my own hopeful skepticism. Scenes from the past echo through my mind, followed by bouts of knee-jerk anxiety. I stop being myself out of fear no one would show up if I did. Guy Smiley returns.

In those moments I remind myself of what I would tell anyone: Cynicism can feel like self-defense, but it's only safe in the way that house arrest is. When queasy thoughts come for me, I use reality testing and behavioral experiments to challenge them. A few years ago, two new professors joined my department. We went out for a drink, and they asked about my experiences. Rather than giving a Smiley spiel about the best job in the world, I told them how Stanford can make anyone feel like they don't deserve to be here, and how often I felt sure the school had made a mistake in hiring me. I can still recall the excruciating seconds of silence that followed—as well as the relief that poured from the two of them right afterward. A polite "getting to know you" turned into hours of authentic sharing. Colleagues turned into friends, and we regularly get together to vent and support one another to this day.

I've slowly come to feel more confident expressing doubts, anxieties, and frustrations. Ironically, this has helped me feel them less, by deepening connections with others instead of performing some positive skit for the world.

Replacing cynicism with wisdom is an emotional climb—away from security and into the unknown, where all our futures lie. Some places, times, and cultures make that path harder and steeper than others.

Chapter 3

Preexisting Conditions

Andreas Leibbrandt wasn't looking for a discovery; he just needed a plane ticket. He was studying for his PhD in economics at the University of Zurich, and his partner was a Brazilian scientist. Leibbrandt wanted to meet her family, but like most students, he didn't have much money. One way to afford the trip to South America was to conduct research there.

Leibbrandt had been studying how organizations shape employees. As a college student, his partner had observed one unusual workplace—a lakeside fishing village in northeastern Brazil. It was the perfect opportunity to fund their travel and they took it. Like many Westerners, Leibbrandt expected Brazilian culture to be friendly and communal and was surprised by what he found at the lake. "The men worked in a very solitary way," he remembers. "They went out at 3:00 a.m. to set their nets and spent the day alone in small boats." The village paths were strewn with garbage, which appeared to bother no one. Fishermen only saw one another when boats vied for the best spots.

The lake was a cutthroat, lonely workplace. How did that affect them? To answer that question, Leibbrandt would need to compare it to another community. The lake flowed into a river. He traveled along its banks and some forty miles later met the sea, where a second fishing village was perched. The lake and ocean communities were similar in size, income, and religion—but not in character. When Leibbrandt arrived at the seaside village, dozens of people emerged from their homes to welcome him. When he got up to leave for the night, two men insisted on guiding him down the treacherous, ten-mile road to town.

What made this second village so much friendlier than the first? Leibbrandt learned that ocean fishing requires large boats and heavy equipment, and as a result: teamwork. Villagers here had to cooperate to make a living. Occasionally, greed would get the best of someone, and they'd go it alone among the heavy swells. That person sometimes wouldn't make it back.

Leibbrandt asked fishermen in each village to play the trust game you tried earlier. As investors, sea fishermen sent about 40 percent of their money to the other player; lake fishermen sent less than 30 percent. As trustees, sea fishermen paid back nearly half of what they got, such that investors earned dividends. Lake fishermen sent back less than a third, so investors lost.

Crucially, residents of these villages didn't start out different from one another. At the beginning of their careers, fishermen in each community trusted equally. But over time, the workplaces changed them. The longer someone worked on the lake, the more his suspicion grew. The longer someone worked on the ocean, the more open and giving he became. By the sea, people learned to trust and were right to do so, winning when they counted on each other. By the lake, they learned cynicism, and were also right, because trust paid off less.

If you plant an orchid in the desert and it wilts, you wouldn't diagnose it with "wilting orchid disease." You'd look to the environment around it. Psychologists like me tend to think about people one at a time. But human beings are products of their surroundings. Brazilian fishermen evolved into different versions of themselves to suit the demands of their social world. Dozens of studies have found that people grow kinder or crueler based on their circumstances.

Cynicism runs in families, but less than half of it can be explained by our genes. Cynics are not born, they're made, and our society is minting them. Modern life, especially in the West, is packed with cultural "preexisting conditions" that tune our instincts toward mistrust and selfishness. We have built a Lake Town of historic proportions. But understanding the forces that breed cynicism also provides hints about how we could do the opposite, building Ocean Villages defined by camaraderie where trust might reemerge.

Inequality

In 1980, the American middle class owned about 50 percent more wealth than the country's richest 1 percent. By 2020, the top 1 percent owned more than the entire middle class. A similar trend, though less pronounced, has pooled wealth at the top in other nations. This has stranded millions of people on the economic margins. Americans born in the 1940s had a 90 percent chance of earning more than their parents; for children born in the 1980s, that chance dropped to 50 percent. Meanwhile, costs of college and property skyrocketed. In 2022, 70 percent of Americans reported being unable to afford the homes and educations their parents enjoyed. Many are just one injury or unexpected hardship away from financial ruin.

Trust binds people into teams, towns, and countries, but inequality dissolves those ties. People in more unequal states and nations tend to be polarized, hostile, stressed, lonely, materialistic, and mistrustful. This isn't because unequal nations are less wealthy. In the US, trust declined during the second half of the twentieth century while prosperity increased. When a smaller share of the population grabs a larger share of resources, cynicism typically follows.

Poorer people have every reason to suspect a culture that leaves them in the cold. But in unequal areas, wealthy people *also* trust less. One reason for this is that inequality creates a *zero-sum* mentality, where one person can gain only if someone else loses. In such settings, even winners are put on edge. Their advantages could be snatched away at any time, and plenty of people would love to do the snatching. Colleagues, neighbors, and strangers become rivals.

Elite Abuse

In East Germany after World War II, the Communist government created the Ministry for State Security, or "Stasi," to monitor the population and root out dissidents. The Stasi, in turn, hired a vast network of informants. Your butcher, drinking buddy, or second cousin could be a poorly paid spy, ready to rat you out at the slightest hint of rebellion. This destroyed trust. As one political scientist describes it, "The very knowledge that the Stasi was there and watching served to atomize society, preventing independent discussion in all but the smallest groups."

In 1989, the Berlin Wall fell, and the Stasi crumbled with it. But their reign left a long civic hangover. Even now, residents of German towns once dominated by the Stasi trust less, vote less, and help strangers less often than people in areas where the Stasi held less power. The secret police are gone, but their ghosts have stolen peace of mind from newer generations.

Thankfully, the Stasi are an extreme example of government abuse. But around the world, people are losing their political voice. The nonpartisan organization Freedom House reports that humanity is experiencing a fifteen-year slide in democratic norms. According to their data, three-quarters of the world lived under a government that became less democratic between 2019 and 2020. US states, too, are steadily disenfranchising people. A political scientist recently calculated each state's "democracy score" by examining factors that increase voters' power—such as letting people register online; and decrease it—such as gerrymandering. He discovered that democracy scores have fallen throughout the twenty-first century, especially in states dominated by Republican lawmakers.

When elites—powerful individuals in government and industry—abuse citizens' trust, that trust disappears. In 2021, UK news outlets reported that Prime Minister Boris Johnson had held boozy festivities while British citizens were in COVID lockdown, unable even to attend their relatives' funerals. Before the "Partygate" scandal, 57 percent of UK citizens felt the country's politicians were "only out for themselves." After, that number ballooned by nine points to 66 percent. The previous nine-point rise in British cynicism had taken seven years; the one before that, forty-two years.

Think back to yourself ten years ago. Since then, has your trust in corporations and government dipped? If so, you're part of a large, jaded majority. You can't blame us, either. In the era of WeWork, Theranos, and FTX, it's easy to assume there are only two kinds of people left: the grifters and the grifted.

That doesn't mean cynical thinking helps anyone. Elite abuses raise pre-disappointment—the assumption that people will do us wrong—to a grand scale, which can hurt victims again. Often, this falls along class, racial, and ethnic lines, and only widens inequality. In 2004, off-duty Milwaukee police officers brutally beat Frank Jude Jr., an unarmed Black man. The following year, 911 calls plummeted, especially in Black neighborhoods. Researchers estimate that twenty-two thousand emergencies across the city were not called in because of the attack on Jude. Black Milwaukeeans had plenty of reasons to mistrust the police, and that in turn left them with fewer defenses against crime.

Pre-disappointment can also worsen health disparities. When the first COVID vaccines were rolled out, millions of people refused them. Some did so because of conspiracy theories. But in Oakland's Fruitvale neighborhood, a largely immigrant community, people distrusted for other reasons. Black and brown patients have long received worse care than white ones, and minority communities might reasonably feel American medicine is not made for them. Fruitvale residents also feared the government. When the pandemic began, the US president had recently vowed to intensify ICE raids and had called out Oakland's mayor, Libby Schaaf, for offering sanctuary to undocumented people. Many worried that vaccine appointments would turn into immigration ambushes.

By August 2021, only about 65 percent of Fruitvale had been vaccinated, compared to over 80 percent in the wealthier area of Piedmont just six miles away. A study of over one thousand Fruitvale residents found that Latinos were a staggering eight times more likely than white people to have positive PCR tests. Across the country, the pandemic produced a three-year drop in Latino life expectancy, far greater than its effects on white Americans. Fruitvale residents had reasons to mistrust authorities, but that mistrust carried its own costs.

A Commodified World

Wealth and corruption among elites can raise everyone else's cynicism. But so can shifts in the way the rest of us live. Imagine doing a medium-sized favor for a friend, like picking them up from the airport at midnight. Would you expect them to repay that kindness and complain if they didn't? If so, you might have an *exchange relationship*, keeping track of the value each person gives to the other. If, on the other hand, you are in a *communal relationship*, you give, take, and share for the sake of it.

Exchange relationships are most at home in the free market. We trust someone will give us the product we pay for or honor a contract because it's in their interest to do so. As the economist Charles Schultze put it, "Market-like arrangements...reduce the need for compassion, patriotism, brotherly love, and cultural solidarity." Schultze meant this as a plug for the virtues of markets, but accidentally named one of their downsides. Trade encourages people to act kindly but makes it harder to know *why* they're being so nice. That shirt really might bring out your eyes, and that joke really might have everyone in stitches, but don't count on the salesperson or waiter to tell you if it doesn't.

Markets are driven by self-interest and make it easy to perceive self-interest everywhere. In my lab, we've found that people who act kindly for the sake of financial rewards—like donating to charity for a tax break—are rated as *more* selfish than people who do nothing at all. Exchange is fine for business, but it muddies altruism.

The refuge from transactional behavior has always been communal relationships, where people stop keeping score and are simply there for one another. But if money is the stuff of exchange, it is kryptonite for communal ties. A date whom you pay for their time isn't a date; they're an escort. Advice you charge for isn't advice; it's consulting. Most friends avoid exchanging cash, but there's a problem: We've started counting the rest of life as though it were money. You can leave the store, but the number of steps you take to walk home, how long you meditate, and exactly how many people approve of your latest social media post are now tallied, bought, and sold.

The psychiatrist Anna Lembke says that "any time we enumerate something, or give it a number, we increase the risk of addiction." Numbers have always been part of commerce, but now they crowd health, approval, and connection. We can call this "market creep," and it changes not only what we strive for, but *how* we strive: by chasing numbers instead of experiences. Quantifying has its upsides, but it can produce a whole new set of anxieties, especially the fear of falling short. Insomnia clinics report that people have begun seeking help not because they *feel* tired, but because smartwatches report their sleep quality isn't up to par.

Market creep changes our relationships to ourselves, and how we perceive one another. Logan Lane, a student at Brooklyn's Edward R. Murrow High School, found the social marketplace suffocating. Each morning, Lane took the Q train to school along with dozens of classmates from across grades. "It was like a runway...," she remembers, "the social media in-person collision." Teens eyed each other and their phones, sometimes scrolling through posts made by peers who were standing just a few feet away.

By then, Lane was already a social media veteran; she'd gotten her first smartphone in sixth grade, along with an Instagram account. She shared goofy selfies along with images of her painting and knitting projects. Her online persona was quirky and ironic, taking part in social media but not taking it seriously. Her life told a different story. Logan scrolled herself to sleep. She "couldn't not post a good picture if [she] had one," and she closely monitored others' reactions.

Most collisions between online and offline life are just fender benders, but some cause real pain. When people record themselves and one another constantly, anyone can be an undercover journalist, and everyone is the story. Against that backdrop, it's hard to know who is authentic, and easy to suspect people are performing. When the pandemic hit and classes moved online, Lane was free to spend every waking hour on social media. Her image of others darkened, and her self-esteem shrank. "I was constantly seeing something better that I could be, someone prettier, someone more artistic," Lane remembers. "I developed this level of shame about who I wasn't."

Social media is entering its third decade of making people feel like this. Facebook was launched at Harvard in 2004, and over the next two years was released slowly across thousands of colleges before its public release in 2006. Researchers recently traveled back in time by examining university records from that period. In the months after Facebook came to a campus, students became more depressed, anxious, exhausted, and eating-disordered. They visited counseling services more often and consumed more psychiatric medicine than before.

One culprit was social comparison. After Facebook arrived, students saw their peers partying, vacationing, and generally sucking the marrow out of life. Social media gives us everybody's highlight reels, making us less satisfied with our jobs, homes, relationships, and bodies. More broadly, though, Facebook and other platforms quantify social life. By counting likes, shares, and streaks, they make it easier to compare, compete, win, and lose—and harder to commune.

In the early 2010s, social media platforms were celebrated as the reinvention of global community. By now it's clear they are markets dressed up as communities, encouraging a transactional form of friendship. A 2018 *New York Times* column offered people a quiz to quantify how useful their friends were, urging them to "identify the people in your life who score highest." We are "marketizing" human connection itself.

If markets have crept into friendship, they have bulldozed their way into romance. At the turn of the twenty-first century, about 5 percent of partners met online. By 2017, nearly 40 percent did, dwarfing any other form of matchmaking. There's nothing wrong with an enthusiastic right-swipe. But apps like Tinder, with their infinite supply and split-second choices, change the way our minds process attraction. Tinder founders modeled their app on slot machines. Love might be a game, but online it's a casino, with each person calculating odds and placing bets. You might date only someone who is taller than five foot ten, has a beard, and is secure in their career. But how tall, how bushy, and how successful is ideal? Swamped with options, users "relationshop," weighing suitors' statistics against one another, like someone comparing TVs before buying. They might also start thinking of *themselves* that way, tuning their lives to stack up.

Billions of us live in Silicon Valley's quantified dream. We are given chances to optimize every facet of our lives, but not warned about how truckloads of data might alter our relationships. Those data, meanwhile, have created unthinkable wealth for the largest companies and richest people on the planet. We are one of the most profitable products in history.

Turning Trust into a Default

History is not a scientific experiment. We can't rerun it a thousand times, tinkering with the world like a set of dominoes to see how different arrangements change their fall. Scientists can't know exactly why trust plummeted in the last half century, but inequality and elite abuse track cynicism across space and time, and all three are on the rise.

None of this means we should wish for rosy times gone by. Infant mortality, famine, and violent death have receded over the centuries. Oppression and injustice still run rampant, but many marginalized communities fared worse just decades ago. According to psychologist Steven Pinker, this is the best time in human history to be alive, and we owe the universe a debt of "cosmic gratitude" for progress that occurred before our births.

Perhaps we do. But twenty-first-century dwellers can't easily compare their lives to the Bronze Age. They *can* see real, troubling trends happening now. It's understandable if instead of cosmic gratitude they—and you—feel worldly dread.

Inequality, elite abuse, and market creep trap us in a global Lake Town, like the competitive fishing community Andreas Leibbrandt studied. What does that do to our minds? The sociologist Émile Durkheim described a modern condition called *anomie*: a breakdown of social values and expectations. Anomie is different from the disappointment you might feel when one person lets you down. It's the sense that society itself has betrayed you. Living in a world that feels more transactional, unfair, and selfish each year can grind away our hope; and, indeed, unequal, corrupt, and commodified settings all raise anomie.

But people don't just respond to culture; we also create it. That level of change can feel distant, but each of us has sway over our social microclimates: schools, families, and neighborhoods. In those places, we can create miniature Ocean Villages where trust is the default.

In 1975, Bill Bruneau was biking the hills of Menlo Park with a tiny Emile in tow when they noticed a crew shooting the science-fiction movie *Escape to Witch Mountain*. The two approached and realized the film set was part of the Peninsula School. Poking around its wild, woodsy campus, Bill knew this was the place for his son. He couldn't afford tuition, but the school offered scholarships to children of staff, which was all Bill needed. He and Emile made the five-mile trip to Peninsula every day. Monday through Friday, Emile attended class. On weekends, father and son worked as janitors.

Peninsula is an unusual community, even by California standards. Its main building is a nineteenth-century Italianate mansion, once white, now tinted by the years into a dusty beige. Its gables and wraparound porch are carved with vine and flower patterns. The place feels like a home because it was one. Most classes are held in one-room cabins surrounded by old-growth forest. In Emile's time, students named and climbed favorite trees. His classmates would shimmy up a three-story-tall cedar called "Flat-top" and eat lunch on its wide branches, their view extending across the bay.

Forty years later, I visited Peninsula to track down clues about Emile's childhood. During my first visit, at least four kids ran by me barefoot, and I later learned that only half of the students wear shoes on any given day. Emile's quirks became noticeably less quirky. The January rain had left enormous puddles between buildings, and elementary schoolers were engineering bucket-boats to row across.

This friendly chaos is by design. Peninsula students take classes in mixed-age groups, mentoring and supporting one another. Children lead discussions from an early age because, as one teacher told me, "The one talking is usually the one learning." Grades are downplayed to reduce zero-sum thinking. In Emile's time, kids invented most of the schoolyard sports, including "ballerina tag" and "four goal soccer." As he remembered, "These games were high on cooperation and creativity, and low on competition and convention."

Peninsula extended Bill's underbearing attentiveness, and planted the seed for who Emile became. His sense of adventure and love of nature grew—classmates called him "the monkey" because of how quickly he scampered up trees. Teachers nurtured students' self-reliance by letting them make their own choices. Research backs their philosophy. When adults (and especially parents) trust children, those kids in turn are more likely to trust their friends, feel less stress at school, and earn better grades.

Linda, Emile's mother, would occasionally appear on campus looking for her son. "Imagine," Emile's wife, Stephanie, reflects, "a homeless, disheveled, and mentally ill woman wandering onto a school campus in the middle of the day, looking to talk to her elementary school child who is not in her custody. Imagine the range of responses that a school could have to this situation, and the way that some responses could make the child ashamed of his mother, prevent mom and child from connecting, or cause trauma."

Teachers chose to welcome Linda, creating a space where she and Emile could be together. As we've seen, adversity can send people reeling or spark their growth, depending on who supports them through it. For Emile, that began with Bill and Linda. Peninsula grew that safety to the size of a school.

Peninsula students were family, and, like siblings, they tussled. When Emile was in first grade, a massive sandbox between classrooms became the battlefield as kids fought over whose structures should outlive the others. Here and in many conflicts that followed, teachers trusted students to mediate their own conflicts. Often, a miniature Emile enacted a miniature version of his future, leading the peacemaking process.

In the 1950s, staff and parents constructed an elaborate tree house in the crook of a stately oak on campus. The climb up is daunting (I was offered a shot at it and quickly declined), and many kids see it as a rite of passage. Three decades after reaching the top, Emile built a tree house like it in his own backyard. When he died, the school added a rope net underneath the tree house in his honor. The net can be climbed, emboldening kids to rise faster, earlier, and more often. If they slip, it catches them safely.

The school's effects clearly rippled throughout Emile's life. His default setting was communal: He gave time, energy, and attention away freely. He caught people when they fell and knew they would do the same for him.

This spirit characterized Emile's work for peace, and also *how* he pursued it. Most scientists enter their profession with the noble mission of uncovering truth, yet many give in to less noble impulses along the way. As a profession, science is breathtakingly unequal: Thousands of students toil, sacrifice, and claw their way toward the promised land of tenured academic posts. Very few make it, and those who do tend to be lucky. (I certainly was.) The upshot is that many good-hearted researchers find themselves pulled into intellectual Lake Towns. Success is measured by comparison; a friend's or colleague's win means less opportunity for you.

A cutthroat culture is stressful for many and bad for science itself, encouraging researchers to hoard information and overhype their work. From the beginning, Emile built his Peace and Conflict Neuroscience Lab to be different, like a nerdy commune. He encouraged students to share ideas and credit. The people on his team didn't strive to be *the* experts on what they studied, but rather to add their brick to the wall of knowledge. "It was refreshing," remembers Samantha Moore-Berg, one of the first scientists to join the lab. "Science didn't have to be this competition, and you didn't have to know all the answers."

This approach might have cost Emile a few accolades, but it also showed the people around him a new way of doing science. After his death, Moore-Berg became the inaugural Emile Bruneau Fellow for Peace and Conflict at the University of Pennsylvania. "His vision has become my vision," she says, "I begin with, 'How can we use science to do good, to create impact?' If I'm ever in doubt, I can just ask 'What would Emile do?'"

A Manual for Ocean Living

From childhood, Emile was lucky to find himself in Ocean Villages. Later he constructed microcultures that reflected these values. How might we take his lead and build our own?

Stop counting: Market creep encourages us to count and compete through life, but the science of well-being is clear: To thrive, human beings need unquantifiable experiences, pursued for their own sake. Like a favorite hobby, communal ties can be blissfully useless. In a marketized world, this means purposefully erasing the way we usually connect, by "uncounting."

If you are part of a class project, car pool, or product team, it's fair to ensure everyone does their part. But with people to whom you are deeply attached, resist the urge to keep score. Regularly, deliberately shred the ledger. This is especially important when helping people. Acts of kindness are one of the fastest, most powerful ways to boost well-being. But the motives behind these actions matter. In my lab, we've found that people become happier and less stressed on days they help friends, *if* they are moved to do so by compassion.

When you do a favor, it's natural to focus on what you're giving—time, sweat, attention—and wonder when payback will arrive. A better option is to think about *why* you're giving: a loved one's need, the affection you have for them, and the difference you can make. Devotion is an antidote to scorekeeping, a side of humanity cynicism can't touch.

Uncounting is relatively easy in some settings, like meditation retreats and puppy love. Others make it nearly impossible. In 2021, Logan Lane was trapped in the quantified social world of Instagram. She didn't need to keep score of who liked her; the apps did that for her. Realizing the toll this took on her well-being, Lane made a choice unthinkable for most American teens: She left, deactivating her accounts and trading her iPhone for a turn-of-the-century flip phone.

Loneliness engulfed her. Classmates complained that she "fell off the face of the earth." Disappearing from social media felt like erasing herself. And yet, someone remained. Lane began waking up earlier and creating more art. Some peers were unwilling to find her offline, but had they really been her friends in the first place? Others FaceTimed on her laptop or— the horror—called by phone. Lane noticed differences in these connections. They were slow and deliberate. They left no public record and offered no clout. They were simpler, and easier to trust. The friends who stuck around were there for the sake of it, and for her.

At an outdoor concert, Lane ran into another high schooler who also had a flip phone. This small miracle sparked an idea. Lane created the "Luddite Club," a group of high schoolers who vowed to reject social technology. Each week, between a half dozen and twenty-five members met in Brooklyn's Prospect Park to read, chat, and generally laze around—away from the digital world's ubiquitous gaze. Their connections are uncounted, unscored, and free.

I admire the Luddite Club and, in true marketized fashion, feel inferior to them, because I've been screen-addicted since Lane was in preschool. My drug of choice is the site formerly known as Twitter, which I joined in 2009. My excuse is that the platform offers chances to learn about new science and share my own work. In fact, my feed is a wintry mix of academics advertising themselves and dunking on one another, people finding creative new things to be angry about, and convulsive news cycles. I pitch in, too, eagerly watching engagement on my posts, sometimes deleting the ones people seem to ignore.

Inspired by Lane, I deactivated my account for two weeks. After clicking the button—which is surprisingly hard to find—I immediately panicked that I'd deleted it for good. Once that was ruled out, slower effects set in. In moments of boredom, I instinctively opened web browsers and typed "Tw...," the same way someone might turn their car toward the office on a Saturday. Events felt different. A Russian oligarch died in an apparent government plot. Presidential candidates debated. I wondered, "What does everyone think about all this?" and realized I would never know. This news cycle and a half dozen more would be over before I logged on again.

But I wouldn't know what everyone thinks anyway. My feed played the role of news but was really just the same few hundred people trying to outyell one another. Then, there was "Twitter-only news": the controversies, villains of the week, and inside jokes that floated through my corner of social media. Others would metabolize these events without me. But what had those stories done for me, in any case? Infighting between psychologists made me lose interest in work; politicians trolling each other lowered my opinion of government. Ex-Twitter dragged me from one crisis to another, like flotsam being pulled by the current. Offline, I regained power over my attention.

My interactions changed, too. During my online fast, Stanford appointed a new provost. Instead of posting about it or watching others announce their opinions, I started a text thread with colleagues. It was quiet and honest, and I trusted they were sharing their real views, more than I would online. A friend released a new album. He probably posted about it and I could have kept up with him from a distance. Instead, I called. He told me all about the experience, and learned that I cared, something that would never have come across if I'd simply watched his life unfold on my screens.

After the two weeks were up, returning to the site felt like chasing a Twinkie with half a pack of cigarettes. Instead of processed sugar and nicotine deploying throughout my body, familiar feuds, characters, and nonstories took over my mind. Most of them were unwelcome, like an annoying uncle you forget between family reunions. I realized how wonderful it was to *not* follow these threads. Fear of missing out morphed into joy. I'm not alone. In 2018, researchers paid nearly three thousand people to shut down their Facebook accounts for four weeks. This digital cleanse lowered people's depression by 25 to 40 percent, comparable with going to therapy.

I still pop onto Twitter to share and spend a few minutes gawking or catching up. But its hold on me is nothing like before, and I try hard to replace life online with regular, uncountable connections. Each time I text instead of lurking or call instead of scrolling, I notice how one-on-one connections bring people into focus and quiet the online world's endless noise.

Play together. Researchers recently surveyed nearly two hundred New York parents about what they wanted their children to believe. A third of parents thought their kids would go further in life if they saw the world as competitive, not cooperative. More than half thought it would help their children to view the world as dangerous. In another set of interviews, 70 percent of parents said they encourage their children to distrust strangers. One mother, who generally trusts people herself, said, "It's crazy, but we have to teach our kids not to help...so unfortunately the first thing to teach my daughter was to be cautious and only talk to known people."

These parents try to keep their kids safe by making them feel unsafe, a strategy that is all too effective. In 2012, only 18 percent of American twelfth graders believed that most people can be trusted, making Generation Z the least trusting on record. The trust deficit might arise because people are growing more cynical, or because older people are being replaced with younger ones—who learn cynicism early.

This hurts kids as they become adults. Psychologist Jer Clifton studies people's "primal beliefs" about life and the world. Primals are surprisingly disconnected from circumstances. Rich individuals don't necessarily think the world is more abundant, and people who live through trauma don't always find the world more dangerous. Of course, the world is safe *and* dangerous, cooperative *and* competitive. But beyond who's right, we can also ask how our beliefs change us. In a study of five thousand people across forty-eight professions, Clifton found that people who thought the world was dangerous and competitive, versus safe and cooperative, had less-successful careers and were less satisfied with their lives.

The first time I visited the Peninsula School, I arrived early and wandered awkwardly, figuring someone would quickly suspect a middle-aged man loitering between the cedars. Instead, kids and adults waved and smiled. Without knowing who I was, a staff member invited me in. Within minutes, I was talking to the school's director. You might think of this place as hopelessly naive. I wondered instead why the rest of us are teaching our kids pre-disappointment. Perhaps that will keep them safer. It will almost certainly shrink their worlds and diminish their trust.

Peninsula stands in resistance to that worldview, and any of us can.

When you lead or care for someone, you *create* their pre-existing conditions. This is especially true for children. Recently, I've taken a closer look at my own parenting. My daughters and I read books and have conversations about how most people are good. But the girls also hear my wife and me griping about politicians and companies. On the way to school, they learn horrendous things about drivers who cut me off.

Awhile back, I began teaching the kids savoring, a practice for noticing the best parts of life. We hold "classes" in ice cream eating, sunset watching, and kite flying. Instead of scarfing down a coconut cone, we eat it slowly, commenting on what's most delicious about it, and how we'd like to remember the experience. After visiting Peninsula, I realized we didn't take time for *social savoring*, slowing down to appreciate everyday human goodness.

So, we started. Now, when I find myself complaining about people in front of the kids, I take pains to also point out positive things others have done. After bemoaning litter in a city park, I tell them about the many volunteers who clean it up. Last week in the car, I got stuck behind a construction vehicle on a crowded street, until another driver stopped so I could switch lanes. Normally, this small kindness would float away in the morning routine. This time, I explained how strangers slow down to help one another.

These instances might feel trivial, and maybe they are. But children are astute scientists, and adult language helps them draw conclusions about the world. In trying to balance the inputs I gave my kids, I began noticing different things, searching for friendly and cooperative strangers, who were not at all hard to find. A habit of speech became a habit of mind.

Trust locally. Connection arises naturally within local communities like Peninsula, and so does trust. A survey of over twenty-five thousand people across twenty-one countries found that only 30 percent believed "most people" can be trusted, but a whopping 65 percent reported that members of their area or village trusted one another. The difference was most striking in cynical nations. For instance, barely 6 percent of Filipinos trust "people," but more than half trust their neighbors.

Cynicism has a neighborhood-shaped hole punched through it, one people can use to create positive change. In Oakland's Fruitvale neighborhood, a well-known nonprofit did just that during the COVID pandemic. For fifty years, the Unity Council has served the neighborhood and especially its poor. It builds, maintains, and runs affordable housing communities, schools, health-care facilities, and senior centers. Unity Council buildings are clustered tightly near the city's main drag. Children go to school next door to where their grandparents meet for chess and Zumba classes. "Most people don't know what we do," says Chris Iglesias, Unity's CEO, "but they trust us."

In Japan, the tight-knit neighborhood of Mano rallied to fight fire together. In Fruitvale, the Unity Council pivoted its long-standing role in the community to fight COVID. The group teamed up with the University of California San Francisco and Clínica de la Raza, another local organization, to train college-age "vaccine ambassadors." These young people learned the science behind both the pandemic and the treatment, and then went block-to-block with iPads, approaching people on sidewalks and at outdoor malls, asking if they were interested in the vaccine and offering to book them appointments.

The Unity Council estimates this campaign helped vaccinate fifteen thousand people in 2021. The community quickly became safer, because citizens were ready to believe the people in whom they already had faith. This same strategy can combat cynicism elsewhere. Researchers find that when conservatives watch Republican politicians encouraging COVID vaccination and upholding the results of the 2020 presidential election, they grow more open to being vaccinated, and more confident in the election themselves.

On our televisions and phones, corruption, inequality, and crime rule. But the grocers, teachers, and friends we see in person show us a version of humanity that is kinder and less suspicious. With the right focus, we can build communities of trust and camaraderie alongside them, our own Ocean Villages that can expand over time.

Chapter 4

Hell Isn't Other People

Without looking anything up, guess the answer to each of these questions:

1. In 2009, the *Toronto Star* ran a social experiment. Their staff dropped twenty wallets around the city. Each contained money and a business card, meaning that whoever found it could contact the owner. How many of the wallets were returned?
2. During the first pandemic years (2020–2022), did people donate to charity, volunteer, and help strangers more, less, or the same compared to the pre-pandemic years of 2017–2019?

Before we reveal the answers to these questions, let's consider why they matter. For millennia, human beings have shared food, shelter, and protection with one another. Cooperation built our species, but it also leaves us vulnerable to cheaters, who take from a group without giving back.

In the animal kingdom, many beautiful decorations are in fact suits of armor. A winter weasel's white fur, a tree frog's Technicolor poison skin, and an oryx's curved horns all help these animals avoid ending their lives as a meal. Humanity's most common predators are other people. In response, evolution equipped us with a mental armadillo shell: *cheater detection*. We naturally seek out signs of deception and malfeasance. This is a great strategy for self-defense, but what protects us in small doses can become toxic in larger ones. People are so vigilant for cheaters that we regularly overestimate how many there are, while ignoring signs of human goodness. Psychologists call this *negativity bias*, and we can test yours by returning to the pop quiz you just took.

1. In Toronto, 16 of 20 (80 percent) wallets were returned. In a large follow-up with over 17,000 "lost" wallets across 40 countries, most were brought back, with return rates reaching 80 percent in several countries.
2. In 2023, the *World Happiness Report*—a global survey that asks about people's experiences and actions each year—revealed that volunteering, donating to charity, and helping strangers all *increased* significantly during the pandemic. For all its horrors, the plague revealed massive wells of human kindness.

Did you underestimate people? If so, you're not alone. No matter how polite Canadians might be, they still guessed that a mere 25 percent of Toronto residents would return lost wallets. And in 2023, I surveyed one thousand Americans, asking them what they thought happened to global kindness during the pandemic. Most reported it had decreased, and barely a quarter noticed the vast COVID kindness that spread across the world. This same sort of mistake is scattered all over our lives. Researchers find that people regularly fail to realize how charitable, trustworthy, and compassionate others are. Helpers are all around; we just don't see them.

Some of these mistakes reflect how our minds are tuned. Negativity bias means people pay more attention to bad things than good. Again, this makes evolutionary sense: It's safe to ignore a sunset, but not a tsunami. As psychologist Fred Bryant says, "Troubles kick our door in and come and find us...whereas the pleasures and the joys, they don't hunt us down and force us to enjoy them. They wait and they sometimes hide."

Negativity bias shapes how we experience the world, and one another. People pay more attention to untrustworthy faces than trustworthy ones and remember suspicious characters more clearly than wholesome ones. When people read about someone who commits both good and bad deeds, they judge that person as immoral, as though his worst actions are the best evidence about his character.

If we think people are worse than they are, we are *sure* they're worse than they used to be. Recently, psychologists reviewed surveys spanning nearly six hundred thousand people, fifty-nine countries, and seven decades. In each, individuals were asked how moral people were—for instance, their kindness, cooperation, and fairness—now, as compared to years past. Most people were breathtakingly negative, reporting decline on more than 80 percent of moral qualities. It didn't matter who they were. Rural and urban, liberal and conservative, baby boomers and Gen Z might not agree on much, but they all believe humanity is in a state of vicious decline. The data beg to differ. A recent meta-study examined cooperation in more than sixty thousand people between 1956 and 2017, and found that people cooperated 9 percent *more* across time, not less. But our misperceptions carry on, making us yearn for a gentler, friendlier past that never was.

The desire to protect ourselves from cheaters is natural and wise. But it can run amok when we cynically underestimate people's virtues. This problem starts in our minds, but it can get much worse when we start to talk—and gossip—with others.

A Worldwide Gossip Engine

In the 1990s, anthropologists stationed themselves in bars, train cars, and university dining halls to eavesdrop on strangers. Two-thirds of the conversations they overheard concerned relationships, social experiences, or other people. In other words, most of the air we exhale while talking is filled with *gossip*. Gossip gets a bad rap. As the famous saying goes, "Great minds discuss ideas; average minds discuss events; small minds discuss people." But research suggests it deserves more credit than that.

As a college sophomore, I moved in with seven other guys. For maybe a week, each person did their part tidying up. But soon, someone began taking advantage. No one could tell why the dishes were piling up in the sink because everyone claimed innocence. Rather than becoming chumps, the rest of us gave up on cleaning. The abandoned food on the counter grew plant life and before long, attracted animal life. Our kitchen devolved into a shameful fiasco.

Without knowing it, we had stumbled into a "public goods" problem, pitting selfishness against cooperation. Scientists create less disgusting versions of this experience online. In a public goods game, four anonymous players contribute money to a common fund, which is then doubled and split evenly between them. Everyone does best if they all contribute, but any one person can earn more by giving nothing and still taking—a form of cheating known as "free riding." In study after study, at least some players free ride. Other players, realizing they can't beat cheaters, join them, until cooperation drops to zero—an economic version of my undergraduate kitchen jungle.

That is, unless people gossip. In another version of the game, players could talk about who contributed and who didn't, and exclude cheaters—voting them off the virtual team. This changed everything. Fearing shame and retribution, people cooperated more, and for longer.

If cheater detection sharpens our minds, gossip sharpens our communities. A person who is bamboozled might not get their money back, but they can spread the word and sanction the villain. But there's a downside: Just like cheater detection, gossip can fall prey to negativity bias. In my lab, groups of four people played a public goods game. Each player could then write a note about someone in their group, to be passed on to future players. Most people didn't free ride. But when someone did, people gossiped about them three times more than they did about fair players. People who read these notes, in turn, wrongly believed free riding was rampant.

Human beings are ravenous for information about one another, but nature has made us crave bitter social calories. Bad behavior grabs our attention, and we talk about what (or who) is on our mind, focusing chatter on the negative. As whispers spread, negative stories multiply and positive ones disappear. This game of telephone leaves people more cynical. In trying to protect their community, gossipers give one another the wrong idea of who's in it.

Gossip started out as a village newscast, spreading slowly among individuals. We still do that, but now, gossip is also piped through the media's global megaphone. Reporters, like everyday gossipers, often act as though their moral mission is to catch cheaters. The journalist David Bornstein says many people in his industry believe "society will get better when we show where it is going wrong." Over the last century, muckraking reporters revealed inhumane conditions in factories and fulfillment centers, the horrors of lynching and police violence, and abuses concealed by the Catholic Church.

If bad news serves the greater good, it also makes for good business, because it feeds negativity bias. A study of over one hundred thousand stories on the site Upworthy found that *each* negative word in a headline increased its number of views by 2 percent. By piling up alarming language, websites can drive much larger increases in traffic. Media companies being companies, they give people what they'll buy. Over the twenty-first century, headlines have featured a steadily growing presence of negative emotions such as disgust, fear, and anger. Even songs have soured. Between 1970 and 2010, the mention of "love" in popular music dropped by 50 percent, while the use of "hate" tripled.

By now, modern media has become a cynicism machine, and it grows more precise each year. When I was a kid, television was a bad-news Pangaea: a single, bleak land where we all watched the same gas shortages, shuttle explosions, and murder trials. But in the 1980s and '90s, cable news stations cultivated loyal viewers by catering to their preexisting beliefs. Our landmass split apart as liberals and conservatives floated onto smaller information continents defined by their own grievances. Three decades later, social media has stranded each of us on an island of one. We click on stories that grab our attention, and algorithms surround us with more of what we fear and loathe.

News outlets cater to negativity bias even when it guides us clearly in the wrong direction. Between 1989 and 2020, Gallup ran twenty-seven polls asking Americans whether the US was experiencing more, less, or as much crime as it had the year before. In all but two of these polls, most people thought crime had increased. In this chart, I capture this through a black line that starts at 0, moving up whenever most Americans reported things were getting worse, down when they reported things were improving, and staying put when they reported no change. This is our shared imagination: a nation steadily devouring itself.

We could scarcely be more wrong. The gray line represents FBI statistics on violent crimes reported per one hundred thousand people in the US. It, too, starts at 0, moving up with increases in crime and down with decreases. Between 1990 and 2020, the actual crime rate decreased by nearly 50 percent.

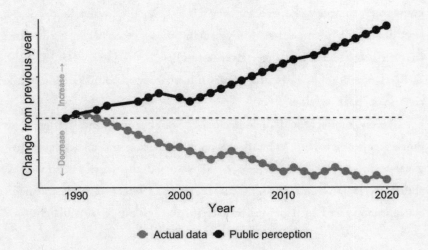

People are safer than they have been in decades but imagine danger behind every corner. The culprit isn't a crime wave, but a "crime-wave" wave. Use of the term "crime wave" in news reports doubled between 2019 and 2021. People who tune in to the news regularly are most likely to believe crime is ascendant, and most likely to be wrong. And it's not just crime. Though research is mixed, much of it suggests that news consumption drives cynical—and incorrect—perceptions of corruption, social division, and economic conditions.

Crime is a very real problem; so are climate change, poverty, and oppression. But when news outlets optimize for eyeballs, they drain our energy. A recent survey asked people to finish this sentence: "The news makes me feel _____." Their answers, which included "hopeless," "agitated," and "despair," cut across politics and identity. David Bornstein compares news media to an ambulance siren sounding every few minutes. "After a while, you'd feel it in your body...you'd be bracing for it all the time." In the past, news happened once each morning in the papers, and once each night on TV. Now it follows us, chiming in our pockets with frequent catastrophes. We have been shocked, and shocked, and shocked again. By now it's difficult to do anything but wait for another jolt.

Talking with Bornstein, I realized how much this cycle has captured me. Like so many people, I brace myself before opening a news app or website, wondering what fresh nightmare will take over my mind. The stories that haunt me most surround climate change and the erosion of democracy, so every screen in my life has learned to feed me just that. An ice sheet the size of Delaware broke off Antarctica. The US has closed nearly seventeen hundred polling places since 2013, many in areas with large minority populations.

Perhaps it's my duty to stay informed, but these stories don't inspire action. They carry me, like riptides, through anxiety, rage, and plenty of cynicism. Sometimes I read late into the night, unable to turn away. Other times, I burn out and stay away from the news for days. So do 42 percent of

Americans, 46 percent of British people, and 54 percent of Brazilians, who in 2022 reported that they actively avoid media.

Balancing Out Badness

Negativity bias and global gossip have worked together to warp humanity's view of itself, like an unfun fun-house mirror. We perceive our species to be crueler, more callous, and less caring than it really is. The philosopher Jean-Paul Sartre once wrote, "Hell is other people." But maybe it's just what we imagine they are.

Most of us are tired of this. A 2021 poll found that close to 80 percent of consumers want media companies to stop shoveling bad news on them. Our exhaustion speaks to what we *do* want: a chance to witness the goodness of other human beings. Thankfully, those positive qualities are everywhere. Opening ourselves to people's best sides doesn't require tuning out our problems. It simply requires balancing our attention through new habits of mind.

Trabian Shorters learned those habits by living them. His hometown of Pontiac, Michigan, a once proud automotive hub, slowly wasted away as he grew up. He was Black and poor, and the world might have never seen him as anything else. But as a child, he tested as intellectually gifted and received a scholarship to attend the Cranbrook School, an elite private institution. Just ten minutes by car from Pontiac, Cranbrook's leafy walkways and stone towers made Shorters feel like he was "off planet." There, he discovered a love of computers that would define his career. After college, he founded a technology company, then changed course and jumped into the world of nonprofits and fundraising.

In coding, Shorters had learned that to hack something, "you have to understand a system well enough to get it to do something it wasn't designed to do." When he transitioned to philanthropy, he saw a system that had stopped working, in part because of how it talked about people. Charities raise money by highlighting the weaknesses of those in need, as though all they can do is suffer and await rescue. Shorters calls this "deficit-framing." To interrupt it, he designed a hack known as *asset-framing*.

Shorters gives the examples of Black and brown children attending poor schools. Media and charities often describe these kids as "at-risk youth" in the thrall of a "school-to-prison pipeline." Educational inequity harms millions of people, but deficit-framing defines children only by the way they are impacted. "It makes it easy to dismiss people who are experiencing a disparity, as if they are the cause," Shorters reflects. Most "at-risk youth" are students, and most students want to graduate. Shorters suggests changing the language we use to reflect this; for instance, by saying: "Students who want to graduate face these obstacles when their schools lack resources." Under this framing, children's goals rise to the surface. Asset-framing doesn't ignore injustice, but refuses to reduce people to helpless victims.

Asset-framing works against negativity bias and plugs into another deep social instinct—to see the good deep inside others. Consider two fathers, Al and Amir. Al used to be a deadbeat dad. In the past, he never showed any real affection for his children or expressed interest in their lives. That's changed, and now he's a caring and involved father. Amir has taken the opposite path: He used to be a caring and involved dad, but now he doesn't want to have much to do with his children.

Both men acted badly, and both acted well. Who are they, really? In a classic study, people encountered several stories like Al's and Amir's, in which someone changed either for better or for worse. Readers were then asked whether the change reflected the character's "true self." When a man changed from deadbeat to wonder-dad, 65 percent of readers thought his true self had emerged. When he went the opposite way, about 70 percent believed his true self had disappeared. In other words, people believe that human beings are good at their core.

Emile would have agreed, but that shouldn't surprise you by now. The twist is that everyone else does, too. Even highly cynical people show this "good true self" effect; it appears in the US, Russia, Singapore, and Colombia. And it reverses negativity bias. How? Especially when we're threatened or stressed, people's worst sides grab our attention and raise our defenses. But when we slow down and feel safe, curiosity rises, and asset-framing comes naturally.

This is one reason it's crucial to think about our local environment. As we've seen, people tend to mistrust humanity in general, but have faith in the people they know and regularly see. In twenty-five of twenty-seven surveys across three decades, most Americans believed national crime was worse than the year before. But in seventeen of twenty-seven, the majority thought crime in *their area* had stayed the same or lessened over the same period. And in the study where people claimed humanity was in a moral tailspin, those same participants said their coworkers, friends, and strangers they interacted with were just as kind as in years past.

Most of us are addicted to the taste of bad news, and producers are concocting potent, bespoke flavors to meet our needs. Simply *knowing* this can help. In 2022, researchers gave about six hundred Americans a menu of headlines to choose from. Some, like "Man Left Seriously Injured After Bowling Ball Attack," focused on single, horrific events—all the ingredients of clickbait. Others, like "Crime Rates in the US Continue to Drop," were positive, broad, and factual. Cynics were more likely to believe crime was rampant, and nearly 70 percent of them opted to read negative stories. Those stories deepened their fear of crime, trapping them in a negative cycle.

But these same researchers discovered an escape hatch. *Before* selecting a story to read, some people read a media literacy message. It explained what you know by now: that people zoom in on negative information, and media companies skew the news to grab our attention and dollars. This message encouraged skepticism, and people who read it were less likely to choose negative stories, especially if they were cynical to begin with.

Clearly, we are starved for better news. But is anyone telling those stories?

Asset-Framing the Media

David Bornstein, like so many journalists, felt stuck by the bad news consuming his industry. That changed when he traveled to Bangladesh to write about its Grameen Bank ("Bank of the Village"). The bank's history began in tragedy, when a famine struck the country in 1974, killing over a million people. Muhammad Yunus, a professor of economics in southern Bangladesh, visited a nearby village and met families on the brink of starvation. Many were skilled craftspeople with a plan to lift themselves out of poverty. All they needed was starter capital, which no one would lend to them. Yunus asked a group of forty-two villagers how much money they would need. They requested $27—not each, in total.

The Grameen Bank was born that day, and over the next three decades it provided microloans to millions of Bangladeshis, almost all women. Grameen's rules are a photonegative of most banks. Instead of requiring collateral, they often lend to people who have *no* assets. Fellow economists thought Yunus was deluded, and that his clients would run away with the money. Their negativity bias was misplaced. Grameen has a repayment rate of 99 percent, comparable to small business loans in the US. As Yunus says, that is the percent of the time he is right to believe in people, and other bankers are wrong not to.

Like Yunus, Bornstein was unprepared for the people Grameen served. "The only images I had of Bangladeshi villagers," he has said, "were people waiting for the Marines to throw them bags of rice after a cyclone." Instead, the villagers he met had "extraordinary agency, were really moving their lives forward."

Bornstein was chastened, and realized some of his bias came from the very product his industry created: deficit-framed media. Like Trabian Shorters's "at-risk youth," the news flattens people affected by poverty, crime, and disaster, characterizing them as simplistic and helpless.

Was there a truer, more three-dimensional way to report? Bornstein and his colleague Tina Rosenberg decided to try. Beginning in 2010, their *New York Times* column, Fixes, reported on "positive deviants"—people and communities doing an unusually good job of tackling important social problems. In other words, they were asset-framing the news. Several years later, a similar project was created by a second, less likely David. In 2016, David Byrne, the artist best known as lead singer of the band Talking Heads, had become worn down by negativity bias. "I wake up in the morning, look at the paper, and go, 'Oh no!'" he wrote. "Often I'm depressed for half the day." Searching for balance, "and possibly as a kind of therapy," he began collecting better stories. This personal stash of good news expanded into *Reasons to Be Cheerful*, an online magazine that covers positive change-makers around the world.

Read both Fixes and *Reasons to Be Cheerful* and you begin to recognize patterns that make asset-framed news different from the usual fare. Mainstream outlets tend to focus on people in power and the lengths they are willing to go to keep it. Asset-framed stories tend to feature everyday people, many of them from underserved communities, helping one another. As Byrne puts it, "Most of the good stuff is local." Fixes reported on Women Overcoming Recidivism Through Hard Work (WORTH), a program created by older inmates in a Connecticut prison to counsel younger ones on trauma, addiction, and résumé writing. *Reasons to Be Cheerful* told the story of a Ukrainian-born Stanford medical student who founded TeleHelp Ukraine, a service that provides free virtual health care to people affected by the war in that country.

These stories—and hundreds of others—show readers people like themselves taking charge in the ways they can, where they can. Instead of sucking people into cynicism, they provide a window into possibility. Bornstein and Rosenberg call their work "solutions journalism." These are not puff pieces about waterskiing kittens, shovels for people who want to bury their heads in the sand. Solutions journalists take national and global problems head-on while centering the dignity and power of citizens to make change from the ground up. Solutions that work in one place can provide a blueprint—and positive pressure—to many others. If one town or state manages to increase college application rates or decrease reincarceration, readers might wonder: "Why can't mine?"

Consumers want more of this. A 2021 poll found that compared to standard, "problem-focused" stories, readers found solutions stories more uplifting, interesting, and fresh. Solutions stories were about 10 percent more likely to cause a change in readers' view of an issue, and 28 percent more likely to cause readers to trust the source they came from.

Solutions journalism remains a minor tide in a sea of cynical clickbait, but its momentum is growing. Bornstein and Rosenberg ended the Fixes column in 2021 to focus on another project, the Solutions Journalism Network (SJN). Since 2013, SJN has trained nearly fifty thousand reporters in asset-framed storytelling. They also host the Solutions Story Tracker, a database of asset-framed coverage of every topic imaginable. If you are looking for revitalizing news, their pantry is stocked. Some of the stories you will encounter in this book are pulled from its shelves.

After speaking with David Bornstein, I decided to change my relationship to the news. Dropping Twitter had helped, but mainstream outlets were still flooding me with negativity-biased stories. I craved balance and sought it out in asset-framed sources. *Reasons to Be Cheerful* is now my home page. Starting my day by learning about positive developments in the world is profoundly uplifting—the opposite of my old news experience.

That doesn't mean I ignore problems. I still read the same news sites as before and think *Oh no!* just like David Byrne does. But now, I take a couple of extra steps. While reading a negative story, I remember that just like most human beings, I'm biased toward badness, and that news companies use that to hook me. This brings on hopeful skepticism. Might there be another side to this story, or at least some reason to hope? Then, I explore the same topic on an asset-framed site. After reading a dismal story about climate change, I searched *Reasons to Be Cheerful* and learned about Connecticut's "green bank," which subsidizes solar panels and other climate-conscious projects, and set the stage for a national green bank beginning in 2023. After reading about voter suppression, I turn to the Solutions Story Tracker and read about a Florida ballot initiative that restored voting rights for previously incarcerated citizens, and was led by them as well.

These stories don't make me think everything *will* be just fine—the false security of optimism. But they jolt me out of the dim stupor news can so easily bring and into a state of hope—knowing that things *could* improve, and that hardworking people are helping all the time.

Any of us can seek out more accurate, less cynical news. But don't forget that the stories *you* tell influence others, too. Gossip is our ancient media, and someone is probably tuned in to yours. Try to balance negative conversation with celebration of the kindness and honesty that you've witnessed. Be someone else's nourishing media.

Chapter 5

Escaping the Cynicism Trap

"We are what we pretend to be, so we must be careful about what we pretend to be."

—Kurt Vonnegut

In 1999, the *Boston Globe* published a scathing exposé on the city's fire department, which was wasting millions of dollars in bloat and corruption. The department's chief resigned in disgrace, and his replacement hired management consultants to trim the fat. Among their targets were firefighters themselves. A report pointed to an "alarming" number of injuries, suggesting firefighters might be abusing the system to sneak time off. Cheater detection inspired a new policy: Now, firefighters hurt in the line of duty would need to go before a doctor to prove they weren't faking and be put on desk duty instead of rest while they healed.

Firefighters loathed this change. Some probably *were* faking injuries, but most risked their lives to serve their communities. As reported in the *Globe*, many "took it as a point of pride that they were tough enough to work through fatigue or illness." Instead, the chief was treating them like teenagers skipping class. "When you're injured, you can't function," said one fireman. Their boss was forcing them to work through pain anyway. Meanwhile, firefighters hadn't received a raise in years. They picketed the mayor's public events, creating "several ugly scenes."

After two years of angry struggle, firefighters and the city finally arrived at an agreement, including a new policy for sick days. Before, sick leave at the department was taken as needed; now each firefighter could take up to fifteen days per year. The chief promised to investigate anyone who abused the policy, and later he would do just that.

The new contract took effect in December 2001 and backfired spectacularly. That year, the entire department had taken about sixty-four hundred sick days. In 2002, it took more than thirteen thousand. Mysterious outbreaks occurred on the Fourth of July, Labor Day, and New Year's Eve. Labor shortages shut down whole stations for days. The number of firefighters who took exactly fifteen sick days increased by almost tenfold. Accused of being selfish, they played the part.

Negativity bias makes us think people are worse than they are, and these mistakes leak into our actions. To stay safe in a selfish world, cynics often take *preemptive strikes*: surveilling, threatening, or harming others. Employees who believe their colleagues are talking behind their back are more likely to eavesdrop on them. Lovers who don't trust their romantic partner tend toward emotional abuse. People who think their friends won't support them through difficult times disappear when others need them.

In sports, the best defense might be a good offense, but in life, preemptive strikes just offend. They are miniature social acts of war, and, like Boston firefighters, people respond by becoming exactly who cynics expect them to be.

Self-Fulfilling Prophecies

Maya Angelou once advised, "When people show you who they are, believe them." But what people show you depends on who *you* are. My students seem pretty interested in psychology when they attend my office hours. I could conclude the world is full of budding social scientists, but it's more likely students save their research questions for me and seldom bring them up with anyone else. Undergraduates also tend to be deferential, sober, and serious with me—qualities appropriate for a professor's office—but which I wouldn't expect them to keep up every Saturday night.

People follow our lead more than we realize, a pattern the psychologist Vanessa Bohns calls "influence neglect." In her studies, Bohns has people make requests of strangers, such as asking to use their phone or for help finding a local landmark. Before they start, Bohns asks participants to guess how many people would comply. It turns out people are terrific at getting others to do things, but terrible at realizing their own power. Participants guessed less than 30 percent of strangers would do what they asked. In reality, more than half did. In another study, people asked strangers to vandalize a library book. Again, they guessed less than a third would comply, and again, more than half did. We underestimate our effects on others' helpfulness, and on their harmfulness.

The stories we tell ourselves about people change how we treat them, which, in turn, can alter the course of their lives. Teachers who think a student is bright will invest more time in that student, who is then likely to thrive. Bosses who take someone under their wing boost that person's chance of success. Friends, colleagues, and neighbors become the people we pretend they are.

Let's return to the "trust game" from chapter 1. Back then, you were the investor, deciding how much of a $10 prize to send to a trustee. Whatever you sent would be tripled, and the trustee could return as much of that money as they wanted. Try to remember how much you sent as an investor.

Now put yourself in the other role, as the trustee. A stranger on the internet chooses how much money to invest in you. Imagine they sent just $1 of the $10. For whatever reason, they decide not to count on you. How would you feel in response, and what would you do? How would you feel if they sent $9, demonstrating a high level of trust?

As it turns out, one person's trust shapes how another person responds. In data from more than twenty-three thousand people across thirty-five countries, researchers found that when investors sent more money, trustees sent back more—not just in dollars, but in percentage. The average investor sent $5, which when tripled left the trustee with $15. The average trustee sent back about 40 percent of that, or $6, leaving the investor with $11, or a profit of $1. If an investor sent $6 instead, the trustee ended up with $18, and sent back about 50 percent, or $9, leaving the investor with $3 of profit. In other words, one dollar of additional trust paid back a 300 percent return. By contrast, the stingier an investor was, the less money trustees repaid.

Why? Cynical investors imagine the trustee will run off with their cash, so they don't send much. But they *do* send a signal, loud and clear: "I don't believe in you." In response, trustees feel betrayed and angry, and they have just one way of repaying that insult—by running off with the investor's cash. When investors trust instead, they send a different message: "I *do* believe in you." Honored in this small way, trustees repay the favor. Economists call this "earned trust": When we set our expectations high, others are more likely to step up and meet them.

These self-fulfilling prophecies play out all over people's lives. When suspicious workers are caught spying on their colleagues, other people become more likely to talk behind their backs. When jealous people accuse their romantic partner of wandering, partners lose interest in the relationship. When people think their friends don't respect them, they pour on snark and irony, making it more likely their friends will actually disrespect them in the days that follow.

Cynics tell a story full of villains and end up living in it. Preemptive strikes also ruin our ability to learn who people *would have been* if we'd treated them better. When someone reacts badly to suspicion and scorn, cynics decide they were right all along, like investigators who claim to catch a criminal after planting evidence on him.

This is what happened with Boston's firefighters. After the chief's preemptive strikes, firefighters took more sick days in response. Reporters noticed the effect, but not its cause. One *Globe* columnist complained, "It's a basic axiom of life that whatever the system, people will find a way to abuse it... Firefighters have shown that their putative pride doesn't make them immune to that impulse."

But that's backwards. People don't always find a way to abuse systems. They fight back when systems and people abuse them. This is yet another way cynicism traps us in cycles of anger and aggression, each person swearing the other started it. The good news is that if preemptive strikes can set off a negative self-fulfilling prophecy, then, by making different choices, we can bend reality the opposite way.

Trust as Power

In 2002, a young FBI agent named Robin Dreeke sat down for a beer with "Ivan," an intelligence officer from a former Soviet nation who was fed up with abuses of power at his agency. Dreeke liked his contacts to be disgruntled, because his job was to convince foreign agents to defect and spy on their own country for the United States. Ivan was interested—the two wouldn't have been talking otherwise. But he would risk his career, family, and life if he took another step. Complicating matters, Ivan didn't know who Dreeke worked for, and who he would be collaborating with if he chose to go forward.

Dreeke was an unlikely spy-catcher. Raised in upstate New York by parents who worked multiple minimum-wage jobs, he chopped wood and dragged it across a frozen lake to heat his home. After school, Dreeke entered the US Marine Corps, where an FBI recruiter asked if he'd be interested in helping his country in a different way. At first, counterintelligence training felt like a tour through the dark arts of subterfuge and trickery. Each spy-catcher then made those ingredients their own. Some played a volume game, blurting out million-dollar offers to dozens of possible sources. Others used deceit or blackmail to paint spies into a corner.

Dreeke quickly noticed limits to these tactics. He lost a source after communicating poorly, making the job about himself instead of the other person. "I was being coy," Dreeke remembers, and it wasn't working. He decided to study with one of the agency's most successful agents, a "Jedi Master" who had single-handedly turned over a dozen high-value sources. This phenom didn't try to hook spies like fish; he talked to them like friends—being radically transparent about who he was and how they could work together. He learned as much as he could about his sources' needs and sincerely asked what he could do to help them.

With Ivan, Dreeke used this tactic. Before the drinks arrived, he casually commented, "The thing about Russian intelligence officers is that they don't realize everyone knows they're in the GRU" (Russia's major intelligence agency). This was a coded signal most agents would be able to read: Dreeke wasn't explicitly outing himself as an FBI agent, but made it clear he was familiar with the world of spycraft. His cards were on the table, and Ivan immediately knew the stakes. He didn't walk away, which Dreeke took as a good sign.

Dreeke started asking questions about how he might be able to help Ivan. Rather than offering a bribe, he connected Ivan to some consulting opportunities, and talked about educational options for his son in the US. The two constructed a relationship over months, then years, bringing the US mountains of vital intelligence. Trust was not a weakness, but a strength, helping Dreeke achieve his aims.

Three decades earlier, Robert Axelrod, a political scientist at the University of Michigan, discovered a similar principle in an entirely different way. Axelrod became obsessed with an old, vexing question about life. If creatures must battle one another for survival, how would we have ever evolved into cooperative beings? Axelrod couldn't re-create the ancient world, so he simulated it through a virtual tournament. Players paired up in the "prisoner's dilemma," a game in which each one decides whether to cooperate or cheat. If both players cooperate, they win more points than if they both cheat. But any player can win even more if they cheat while their partner cooperates—like a prisoner who gets off the hook by ratting out their partner in crime.

The players in Axelrod's tournament were not people, but old-fashioned AI programs. Mathematicians, economists, and psychologists from all over the world mailed in "contestants": lines of code specifying how their agent would play the game. Some were relatively naive, cooperating even when their partner didn't. Others cheated to see what they could get away with. Axelrod let these agents loose like windup toys. Each played several rounds of the prisoner's dilemma with every other player. Players who ended up with more points then "reproduced," creating more of themselves like an animal successfully passing on its genes. The new generation then played each other, and the process repeated. Axelrod's tournament was a tiny microcosm of evolution.

Most programs included many lines of code: rules within rules and contingency plans. Most turned out to be too clever by half. The winning program, called Tit for Tat, was the simplest. It began by cooperating, and afterward did whatever the other player had done on the last turn. If you messed with it, you got messed with. If you cooperated with it, it cooperated back.

Tit for Tat was a perfect skeptic, first learning about its partners and then acting on what it found out. It played great defense against cheaters and built alliances with kinder programs. But it had one Achilles' heel. If its partner cheated once, it would cheat back, and both players would end up cheating on every turn, trapped in a losing strategy. The fix, and eventual champion of future tournaments, was Generous Tit for Tat, or GTFT. GTFT reciprocated most of the time, but with a twist. Occasionally, it would cooperate even when the other player cheated, turning the other cheek. The program gave its partner chances for redemption. It added a dash of hope to its skepticism. This sprang GTFT and its partners out of the cheating trap.

Axelrod was bemused. "Surprisingly," he wrote, "there is a single property which distinguishes the relatively high-scoring entries from the relatively low-scoring entities. This is the property of *being nice*." By now, this shouldn't surprise you. GTFT wasn't just friendly for the sake of it. Its niceness was a wise tactic, which enabled it to thrive and multiply.

That doesn't mean we should trust everyone, all the time. If in Axelrod's tournament, agents had played only one round of the game with each other, cheaters would win every time. GTFT won because agents had to work together repeatedly. Cheating once could win a player some points at first, but it'd cost them many more over time as their partner cheated in return. "For cooperation to prove stable," Axelrod wrote, "the future must cast a sufficiently large shadow."

In other words, trust is most powerful in long relationships. This is also where mistrust is most poisonous. Dreeke witnessed this in the world of intelligence. A spy-catcher might use trickery to tease out bits of information, but once they were found out, their source would walk away. According to Dreeke, manipulation "is a weapon that will always blow back on you eventually."

You can safely ignore the email informing you that a prince has left you $12 million, and the influencer promising that one weird trick will give you passive income and clear up your skin. The person who's taken advantage of you time and time again can be kicked to the curb. But if you're trying to build or strengthen a connection, trust isn't gullibility. It's a form of power, one that builds connections, creates new opportunities, and changes people for the better.

In our cynical moments, it's hard to remember this. We've all gone out on a limb for someone and watched as they snapped it behind us. Pre-disappointment is a state of mind, the belief that everyone will let you down. Preemptive strikes are the actions we take in response. Both feel safe, all the while trapping us in lonely, bitter versions of our lives. Escaping requires us to expand our thinking about how people work.

A Reciprocity Mindset

In 2021—after Emile died—his wife, Stephanie, found a note on his phone outlining a lecture on the neuroscience of racism. The bullet points read like a prose poem about the brain:

- Change is a central organizing principle of neuroscience (and Buddhism too!).
- Not that we are able to change, but that's how our brains are fundamentally designed.
- In grad school, my research team discovered that synapses—the connections between nerves and their targets—are incredibly dynamic, driving home to me that at the sub-cellular level, our brains are made to change.

Here's another way of putting this: Biologically speaking, the only thing human beings can't do is *not* change. The qualities we think of as constant—personality, intelligence, and values—evolve over time, right along with our brains. This can be unsettling, or empowering. The ship of your life is sailing. You can't stop it but can steer its path.

The cynic in each of us typecasts others based on their worst actions, believing someone who cheats will always be a cheater. If you see the world this way, you might treat new people like slot machines, trying to figure out which ones will pay off and which ones will steal your coins. But as we've seen, people don't just change; *we* change them through our expectations and actions.

With great power comes great responsibility, but with *any* power comes *some* responsibility. Recently, my lab tested whether teaching people about their power would help them wield it more carefully. Half of the participants in our experiments learned a fixed mindset about trust: They read that, like slot machines, some people pay back investments, and others don't. The other half learned a "reciprocity mindset": They read that people are more likely to repay investments when others believe in them, and more likely to cheat when they're treated like cheaters.

After they learned one of these mindsets, we asked these people to share stories of trust from their own lives. People who learned a fixed view remembered times they'd been burned. "I had a relationship with someone who proved untrustworthy," one wrote. "He is not likely to change. Not worthy of my continued hope."

People who learned a reciprocity mindset wrote about their influence over others. One described their eight-year-old son: "I've always told him that my trust is his to lose...after having that conversation he's become much more open to me." Another wrote, "I decided to let my current boyfriend in despite being absolutely traumatized before and I'm [now] in the healthiest and happiest relationship of my life."

A reciprocity mindset also changed what people did next. Given a chance to play the trust game, they invested more in strangers. This was nice to trustees, who felt happier, more respected, and closer to reciprocity-minded investors. It was also "nice" in Axelrod's sense: a wise strategy for success. Trustees who received larger investments paid more back to the people who'd sent them—a classic display of earned trust.

In other words, when people learned how powerful trust can be, they used it, like Robin Dreeke and Generous Tit for Tat. Changing their beliefs changed their reality, improving outcomes for them and their partners. In the lab, this meant making more money in a trust game. Elsewhere, it could matter much more. In the right hands, a reciprocity mindset could help people build stronger connections, brick by brick.

Leaps of Faith

As Emile reached adolescence, Bill and his new wife moved the family to Willits, a forest hamlet of less than five thousand people located two hundred miles north of Stanford. Emile enrolled in seventh grade there and quickly realized how special his former school, Peninsula, had been. At Willits Middle School, teachers were more controlling, and kids sized each other up. For the first time, Emile felt shy and self-conscious. He managed these emotions by becoming physically strong. When Emile visited his old friends at Peninsula, one marveled that he "just showed up one day buff."

Emile wrestled, pole-vaulted, and sprinted his way through high school, and at Stanford joined the men's rugby team. Rugby is a bruising sport, with an injury rate three times higher than American football. Emile was a fierce player and vicious tackler. In a game against Sacramento State, he got into a dozen high-speed collisions in a row. His coach, Franck Boivert, remembers the moment in disbelief. "He was getting hurt, hurting other people, and the whole time was on cloud nine from the sheer intensity."

Despite its brutality, rugby is a deeply cooperative game. In football, as someone runs with the ball, their teammates block rival players out of the way. Rugby prohibits blocking; instead, teammates run behind the player who has the ball so he can throw it to them if he gets in trouble. "You don't remove obstacles from your teammate's path," Emile once wrote, "but you instead are there for them when they need you. This is called being 'in support,' and you typically let them know you're there by saying 'I'm with you.'"

Emile gravitated toward the game's communal nature. So did Coach Boivert. Before matches on Stanford's Maloney Field, he would gather the group in an equipment shack, where they would psych themselves up with chants and songs. Boivert's favorite, from his native Catalonia, was "Bon voyage to the warriors. To their people they are faithful." The team was faithful to one another for life. They attended each other's weddings and watched one another's children. Emile was buried in his rugby uniform. Teammates wore theirs to the funeral, and those who couldn't join took pictures of themselves in theirs from around the world in tribute.

Boivert taught Stanford "ruggers" to have faith in one another, and showed his own faith in them as well. Compared to other teams, his players were small, inexperienced, and nerdy. Some coaches might have micromanaged them, drilling every action and move, worrying they would get beaten otherwise. Boivert took the opposite approach—underbearing attentiveness. In practice, he dropped the team into gamelike situations and just let them play.

This made each session more fun, which for Boivert is a tactic. "If you don't have fun," he tells me, "your level of attention drops. If you have fun, you're always alert." It also signaled trust in his players. "Each person must bring their own answer to the situation they're facing, and take their own initiative," Boivert says. His belief in players shone through even after failures. Following a demoralizing loss, he surprised the team with a wine and cheese celebration. This unwavering confidence drove players to earn his trust, and brought them to heights they had no business reaching. The team traveled the world, beating larger men and more-established clubs.

After college, Emile returned to Stanford to coach the women's rugby team, an opportunity he relished. Like his own college squad, the Stanford women were small, scrappy, and not very respected. Emile wrote that they were "perhaps better described as a social club with a rugby focus." To take them to the next level, he adopted Boivert's approach: focusing on spontaneity and pouring trust into his players.

One of his stars was Janet Lewis, who played fly half. Like a quarterback in American football, the fly half orchestrates their team's offense. Janet had been on teams where coaches imposed order on players. "I knew how to follow recipes," she remembers. Emile refused to give her any. Where other coaches would bark from the sidelines, Emile watched practice carefully but stayed noticeably quiet. "That left room for us to relax and play," Janet says, "to find our own strengths and lessons from the game."

Emile stuck to this underbearing attentiveness as the stakes rose. Nervous at the beginning of the season, Janet asked Emile to go over strategy. How would she know when to run which play? Could he give her a set of instructions? Smiling, he advised her, "Just read the field." She was incredulous, and asked again, rephrasing the question. But he just repeated his answer. "We would practice a lot to build my instincts... but he wanted [the game plan] to come from me," she recalls.

Like Franck, Emile showed confidence in his people, which built their faith in themselves. Under his guidance, the team won a regional championship and qualified for the national competition. The next year, they won that, too—the first of four national championships for the Stanford women so far.

Ernest Hemingway once said, "The best way to find out if you can trust somebody is to trust them." He was half right. Trust doesn't just teach us about people; it *changes* them. It's a gift they repay.

A reciprocity mindset means understanding this. *Leaps of faith* are actions inspired by that knowledge: deliberate bets on other people. Where preemptive strikes bring out the worst in others, leaps of faith bring out their best—especially when people feel our belief in them, like Emile felt from Boivert, and Janet Lewis felt from Emile. Trust is most powerful when people offer it loudly: clearly giving others the chance to show us who they are.

Loud trust can look irrational. Someone loans their car to a new friend; a manager delegates a sensitive task to a junior employee; online daters fly across the country to meet up. But it's precisely that rash, uncounting quality that makes trust most powerful. In laboratory studies, people who invest in others in "uncalculating" ways—quickly or without knowing the odds of being repaid—are more likely to inspire others to earn that trust.

That doesn't mean it comes naturally. Our cynical minds play movies of betrayal on repeat. Trust is powerful, but also frightening, especially for those of us who have been burned before. But leaps of faith come in all sizes. We can start small, loaning our friend a bike instead of a car, or giving a new employee a less critical task at first. Mixing loud trust with a dose of skepticism allows us to build relationships warmly and wisely. This is how Robin Dreeke began his work with Ivan. It's how Generous Tit for Tat won Robert Axelrod's computer tournament. It's how Grameen Bank launched its support of entrepreneurs in Bangladesh.

As we turn up the volume on trust, amazing things happen. Loud trust might even have helped prevent a war. In June 1963, eight months after the Cuban Missile Crisis, the US and the USSR were each amassing nuclear weapons, afraid of being outgunned by the other. Then, in a speech at American University, John F. Kennedy made a radical move: declaring peace. Kennedy rejected the cynical way that both nations had given up on hope:

> First: Let us examine our attitude toward peace itself. Too many of us think it is impossible. Too many think it unreal. But that is a dangerous, defeatist belief. It leads to the conclusion that war is inevitable—that mankind is doomed—that we are gripped by forces we cannot control. We need not accept that view. Our problems are manmade—therefore, they can be solved by man.

JFK announced that the US would stop nuclear testing in the atmosphere. This was a unilateral move with no assurance the Soviets would follow. Hawks in Kennedy's administration accused him of showing weakness. But Nikita Khrushchev, JFK's counterpart in the USSR, repaid this openness. Soviet citizens were usually barred from hearing any American words, but JFK's speech was rebroadcast around the country. Khrushchev then declared that the USSR would stop producing nuclear bombers.

De-escalation happened one step at a time: a nuclear test ban, the reopening of trade between the nations, even talk of exploring space together. It was a game of generous tit for tat on a global stage. Both nations trusted loudly and took small but growing leaps of faith, until the shadow of war—at least for a time—receded.

Spies, computer programs, rugby players, and heads of state don't have much in common. Still, they share a lesson: Cynical stories can become self-fulfilling, but hopeful ones can, too. By feeling—and wielding—the power of trust instead, we can transform cynicism's vicious cycle into a virtuous one.

Section II

REDISCOVERING ONE ANOTHER

Chapter 6

The (Social) Water Is Just Fine

By the time Atsushi Watanabe realized what was happening, he'd been in his room for nearly six months. Japan celebrates Umi no Hi, or "Sea Day," in July. On New Year's Eve, Watanabe was watching an online streaming site—the same one on which he lurked many waking hours. Another viewer commented, "The last time I saw the sky was Sea Day." Watanabe realized, with horror, that was true of him, too. He had joined the *hikikomori*, a Japanese term for people living in total social isolation.

How had he gotten there? In a series of email conversations, Watanabe shared his story with me. He grew up in Yokohama, a port city about twenty-five miles south of Tokyo. It was a chaotic home—Watanabe experienced bitter fights with his older sister and harsh criticism from his father. Atsushi found escape in television, and self-esteem in art. From a young age, he had displayed a gift for drawing and crafts, and he eventually attended a specialized high school followed by Tokyo University of the Arts.

Setting out for college, Watanabe discovered the art world could be just as cutthroat as any other profession. Young creatives felt constant pressure to break new ground. International markets kept rankings of artists' quality and popularity. Exhibits, press clippings, and awards were tallied and compared. Harassment was rampant and often accepted as the status quo. Instead of an Ocean Village, Watanabe found himself in a creative Lake Town. He had struggled with depression and anxiety in the past, but at university they took over his mind. His world seemed full of indifferent, greedy fakes. He smashed his phone and stayed home for longer and longer hours, drawing inward. Disappointment hardened into pre-disappointment.

Watanabe retreated to his childhood home but found little comfort there. His father remained austere and critical, and his mother appeared unwilling or unable to intervene. His resentment toward both of them swelled. Soon, the only place he felt safe was in his room. So, he stayed there, eventually surrounded by food and bottles of urine, as the weeks ticked by.

Hikikomori translates to "pulling inward" and describes adults who live in total isolation for at least six months. National surveys suggest that about one in every hundred Japanese adults live this way, but it is not unique to that country. Cases have been reported in Spain, Oman, and the US, and new research suggests that nearly 1 percent of adults in many nations live in near-total isolation. The *hikikomori* suffer extreme withdrawal, but milder forms of loneliness are rampant. In 1990, just 3 percent of American men reported having zero close friends. In 2021, 15 percent couldn't name any—a fivefold increase in just two decades. The trend was nearly as sharp for women. Adolescent loneliness is climbing even faster. In a survey spanning thirty-seven countries, nearly twice as many teens reported feeling lonely in 2018 compared to just six years before. And that was before the pandemic.

Loneliness intensifies depression, disrupts sleep, quickens cellular aging, and makes it harder to bounce back from stress. It even worsens the common cold. In one unpalatable study, researchers pumped a spray of rhinovirus—which causes mild respiratory infections—directly up people's noses. Over the next week, whenever these poor subjects blew their noses, a lab tech would weigh their tissues. Lonely people caught colds more often and produced more snot than their better-connected peers. And, like cynics, lonely people tend to die earlier. In one mega-study of more than three hundred thousand older adults, severe loneliness increased mortality risk as much as smoking fifteen cigarettes a day, drinking heavily, or not exercising.

In your twilight years, the science suggests, you might be better off boozing, smoking, and carousing into the night with friends and family rather than drinking tea and power walking alone.

Loneliness is a neurotoxin, and it's spreading. In 2023, US surgeon general Vivek Murthy issued a national advisory on what he calls an "epidemic of loneliness and isolation," warning that if we fail to build stronger social connections, the country "will pay an ever-increasing price in the form of our individual and collective health."

There are dozens of causes for modern loneliness, just as there are many reasons people feel alienated at work and hateful in their politics. But as we'll see in the chapters to come, cynicism plays an underappreciated role in all of them, an invisible thread tying together vast and varied problems. Thankfully, once we notice it, we can start to unwind that thread.

Social Shark Attacks

I grew up in Massachusetts, and on the occasional summer Saturday my mother would drive her mother and me to Cape Cod for a day at the ocean. The ladies would stay ashore; I would swim out as far as I could until beachgoers looked like small, fuzzy blobs. Far from anyone, I'd wile away afternoons punching through the waves. The water churned, and I felt utter peace. During a difficult stretch of my childhood, the sea was a respite.

That is, until the shark attack. It didn't happen to me, but ten thousand miles away, where a great white chomped an Australian nearly in half. Please be assured: You and I are around fifty times less likely to die by shark bite than by lightning strike. And yet, the news report's gruesome details lodged in my eleven-year-old mind. In dreams, I'd find myself falling from a great height into a pitch-black ocean, dull eyes appearing under the surface as the water rose toward me. My next time swimming, a fin popped through the froth between waves. At least I swore it did. I started hugging the shore more closely, and within months barely entered the water at all.

Imaginary shark attacks took away my real joy. For many of us, talking with other people can be like that. Across dozens of studies, researchers have compared social predictions to reality. Some participants in these experiments were asked to *imagine* interactions with other people, while others *actually* interacted with people and reported back. In virtually every case, people's expectations were worse than reality. Chicago and London commuters reported that striking up a conversation with another passenger would be dreadful; less than 25 percent said they'd give it a try on their own. But when scientists instructed commuters to talk with a stranger anyway, many said it was the best ten minutes of their day. They and their new acquaintance vented about the weather, discovered common interests, and sometimes even ended up friends.

We're wrong about strangers, and about non-strangers. People say they would prefer small talk with acquaintances, worried that "large talk" about important, emotional subjects will be too heavy. In reality, deeper conversations leave both parties more fulfilled. People predict that asking for a favor will put their friends out, but others are usually glad to help. We imagine our compliments, gratitude, and support will fall flat. In fact, they boost people's mood and draw us closer.

Relationships falter when people don't realize how much language can harm; cast-off insults by a parent, friend, or lover slip under our skin like a splinter. But how many relationships stall because we forget the *good* our words can do? We keep that conversation on the surface, or don't send that text, sure that's what everyone wants, even when leaning in could make a new friend or save an old one. Your parents might have advised that if you don't have anything nice to say, you shouldn't say anything at all. But we might consider another suggestion: If you do have something nice to say, just spit it out.

Interacting with others is more pleasant and meaningful than most people realize. That's the good news. The bad news is that it's hard to understand this, in part because of negativity bias. When we imagine chatting up a stranger, our minds flood with worst-case scenarios: rejections, glacial pauses, someone rolling their eyes and putting their headphones back on. When we envision saying things we've been meaning to say to people in our lives, we might still picture disaster. These are social shark attacks: dangers that are much rarer than we realize, but so frightening that they dominate our imagination.

As a secret introvert, I live this tension between expectation and reality. Before and after public lectures, I prefer time alone. I commonly break into a cold sweat after running into an acquaintance, then worry they can see me sweating, which makes me sweat more. I try not to cancel plans but gleefully celebrate when others do. And yet after being with people, I feel differently. There are many times when—hours before a party or dinner—I would pay embarrassing amounts of money to stay home, but hours after, I'm glad to have seen friends.

For me, going out is a little like working out: It sounds awful before, but feels great during and after. This appears true for other introverts. Across several studies, people were assigned to act as if they were extroverted—outgoing, talkative, and assertive; or introverted—quieter and more passive for hours or days. Those who performed extroversion reported being happier, even if they were introverts by default.

Social shark attacks might reflect personal insecurity—maybe my jokes are dreadful, I have halitosis, or I'm a burden on my friends. But oftentimes, they hide our negative view of others. Atsushi Watanabe's parents betrayed him, so he saw betrayal everywhere. Alan Teo, a psychiatrist who studies *hikikomori*, describes similar judgments in other patients. "There's this perception that people are wronging me," he says, "a suspicion and rejection that feels well out of proportion with reality."

A subtler cynicism happens in my own academic backyard. Each fall, I teach Introductory Psychology to hundreds of Stanford students, and office hours with them give me a regular pulse of campus life. In 2020, the class went virtual during the pandemic. Over Zoom conversations, students told me how much they hungered to return to campus and their friends. In 2021, they did come back, but something was different. The dorms seemed quieter. Students complained that it was harder than before to meet people or even connect with peers they'd known pre-pandemic.

I wondered if undergraduates might have the wrong ideas about their peers, and if that might be interfering with their social lives. So, in 2022, my lab surveyed thousands of students using two sorts of questions. The first asked them about themselves. How much did they care about their peers, enjoy helping other people, and want to connect with students they didn't know? The second asked about their perceptions of the average Stanford student.

We discovered not one, but two Stanfords. One was *real*, made up of students' self-reports. This campus was extraordinarily warm. Eighty-five percent of students said they wanted to meet new friends. Ninety-five percent said they enjoyed helping peers who were feeling down. Their empathy was through the roof. The second Stanford was a brittle, prickly place located in students' minds. They believed their "average" peer was relatively unfriendly, judgmental, and callous.

My students aren't alone. I've surveyed school systems, government officials, and international companies and the same story emerges almost every time. The average person in each group is empathic and interested in supporting others. The *imagined* average is less kind, more competitive, or downright toxic.

People are wrong about social interactions because they underestimate each other. Like other forms of negativity bias, this one comes with its own actions—not preemptive strikes, but preemptive retreat. Afraid of the unfriendly people in our minds, we avoid real ones. At Stanford, the less kind a student *thought* their peers were, the less willing they were to disclose their struggles to friends or strike up a conversation with classmates. The less they risked, the less they were able to fact-check their fears and realize how many caring, open-minded people were all around them.

Young adults around the world are experiencing a surge of anxiety, depression, eating disorders, and self-harm. One culprit is isolation, which can in turn reflects an underlying cynicism—the idea that others don't want or need us.

Wrong Diagnosis; Wrong Prescription

Isolation wears us down quietly, and when people feel its consequences, they often blame other causes. Lonely individuals tend to visit primary care doctors and emergency rooms with physical complaints. Medicine specializes in bodies, not communities, so most doctors prescribe pills or nothing at all.

Our culture often prescribes even more solitude. Consider burnout, first described decades ago in nurses whose work left them emotionally exhausted. Burnout plagues caring professions like medicine and teaching but has crept up among the rest of us. Depending on the study, between 20 and 50 percent of college students report some burnout, and between 5 and 10 percent of parents report *severe* childcare burnout. During the pandemic, burnout symptoms spiked among essential workers, parents, and the population at large.

Lots of people do not feel like they have just begun burning out. They are spent charcoal, barely able to remember any time before they were set aflame. More of us than ever are investing in one big fix: *self-care*, activities meant to distance people from their troubles. Early in the pandemic, Google searches for "self-care" more than doubled. Companies, schools, and hospitals dedicated days for people to decompress. And even before all this growth, the self-care industry had already reached over $10 billion per year in revenue.

Binge-watching, bonbons, and bubble baths are terrific (especially together), but sometimes self-care is an answer to the wrong question. Burnout's scientific godmother, Christina Maslach, describes this problem as having multiple dimensions. Burned-out people feel things they don't want, like distress and irritability. They also lose things most of us *do* want, like a sense of purpose. And in their exhaustion, burned-out people grow more cynical, tuned in to others' selfishness. When you have nothing left to give, everyone seems to want something.

Exhaustion, distress, loss of meaning, and cynicism are all symptoms of burnout, but they come from different causes. Being overworked cranks up exhaustion; a toxic work environment does the same for cynicism. They also have different solutions. Self-care protects against distress and exhaustion, but it might not bring back people's sense of purpose. A better way to do that is to be there for others. When people give time, money, and energy away, they often feel replenished. Research finds that volunteers who counsel strangers through their problems feel less depressed, and students feel less lonely on days they help a peer. In a recent study of health-care workers, the *only* factor that decreased cynicism was compassion toward others, not self-care.

Helping others is a gift to ourselves. But again, people tend to ignore the good news, consistently—and wrongly—predicting they would be happier spending time and energy on themselves. We see others as more selfish than they really are, and also project that same cynicism into the mirror. This leads us to make social errors. In one sad but telling study, college students reported each week on how they felt, and on their social goals. The more anxiety and depression they felt, the more they focused on themselves. The more they focused on themselves, the worse their depression became.

Most doctors don't know how to diagnose disconnection, but neither do most people. Languishing alone, many of us decide the best medicine is to be alone in nicer ways. Companies are more than happy to provide us with pricey products for doing just that. In need of community, we are corralled back into a market, dragging us further apart.

As we emerge from the pandemic, millions of people, especially in younger generations, hope to work from home for the rest of their lives. Seamless, a food-delivery app, recently plastered New York City subways with ads that read "Over 8 million people in New York City and we help you avoid them all."

Solitude is having a well-deserved moment. It provides space for creativity and peace. But flourishing is often out there, with everyone else. Being alone feels easy, but after a while, it makes getting together harder. Inertia takes over; the social sharks grow bigger and toothier.

Escaping Aloneness

After spending more than seven months in total isolation, Watanabe began to think he would die without leaving his room. Then, suddenly, the choice was no longer his. His father contracted a company that specializes in dragging *hikikomori* from their homes and into asylums. Upon learning this, Watanabe exploded with rage, broke down his door, and stormed into the unfamiliar living room light. The coffee table was cluttered with books he'd never seen. Flipping through them, Watanabe realized his mother had bought manuals for family members to learn about and help *hikikomori*. "She was trying, in her own way, to get to know my mind," he remembers.

Confronted by this love, Atsushi's defenses toppled. When his mother returned, the two talked for hours, finally opening up about their pain, failures, and what they could salvage. His throat dried up several times from lack of use. It would be months before Watanabe would return to the outside world, but the conversation changed him. "That day, the captivity of my heart began to slip away," he says.

Like Megan, the onetime QAnon believer, Watanabe was freed by the safety of a relationship, which gave him the foundation to free himself further. It would not be easy. Watanabe's mind and body had changed during his time alone. His clothes no longer fit; his hours were nocturnal. With his mother's support, Watanabe checked himself into a clinic where he lived for three months, practicing yoga and pottery and rebuilding his health.

At the clinic, a doctor introduced him to Naikan, a form of self-reflection grounded in Japanese Buddhism. Translated as "seeing oneself," Naikan begins with the idea that our minds trick us into unfair judgments. As a remedy, practitioners take part in a careful accounting of their relationships through questions such as "What have I received from others today?" and "What have I given to others today?"

Many of us ignore what people do for us and zoom in on the indignities we suffer. During conflict, we have photographic memories of others' wrongdoing, and convenient amnesia regarding our own. Like reality testing, the therapeutic practice of questioning our beliefs, Naikan replaces negative assumptions with hopeful skepticism.

Watanabe, a self-described "black-and-white thinker," had seen both his family and culture as rigid and intolerant. Through Naikan, he learned to consider other perspectives. He felt not just his pain, but his mother's. He witnessed not just how unfair the art world could be, but also the many people working to improve it. "The difficulty I had in living almost disappeared," he reflects.

Reality testing revolutionized Watanabe's well-being. At Stanford, we're using this approach to fight social shark attacks. In 2022, we launched an advertising campaign. The audience was Stanford undergraduates. The product was also Stanford undergraduates. We knew the average student was wrong about the average student—and we had data on what they were really like. Through a series of campus-wide conversations, we showed students to themselves, teaching them that most of their peers are curious and kind.

Misperceptions like the ones we saw at Stanford are everywhere, meaning skepticism and data can help people almost anywhere rethink their assumptions. But the only way to truly test the social water is to jump in. Despite our fears, it's usually just fine. Psychologists recently invited over three hundred people to take part in a "conversation scavenger hunt." For a week, they were challenged to talk with strangers of all types, for instance, someone who had eye-catching hair, was wearing a scarf, or looked artistic. Before starting the hunt, players estimated they would have to *try* to talk with an average of two strangers to complete each challenge. In other words, they expected 50 percent of people to reject them. After racking up experience, their expectation changed: Now they anticipated that about 80 percent of people would want to chat. Collecting real-world data made them realize how open others really are.

These leaps of faith recalibrate our minds. Can they fight loneliness? In the UK, physicians have begun a practice known as "social prescribing." After addressing a person's aches and pains, they ask about their relationships and offer fixes for what's missing. A doctor might "prescribe" an avid biker to join a cycling club. Bookworms are matched with volunteer positions at a local library.

Early in the COVID pandemic, the British National Health Service (NHS) devoted more than ten million pounds to expand social prescribing. More research is needed to understand its effects, but the initial evidence is promising. People who are given social prescriptions report an increased sense of connection, meaning, and well-being. You might think this comes at the cost of burdening doctors—on top of medicine, they now have to handle social work, too? In fact, these programs appear to make treatment more efficient. One early study found that social prescribing decreased patients' loneliness, and in turn dropped their visits to primary and urgent care facilities by as much as 50 percent.

In his 2023 advisory, the US surgeon general urged our nation to follow suit, training health-care professionals in social wellness, expanding community groups, and cultivating a "culture of connection" in which people regularly practice kindness. These are wonderful goals. But even if we prescribe social medicines, people might not take them if they don't have faith in one another.

I don't know your social world, but if you're like most people, you probably don't realize how much the people around you want to talk and how fulfilling it will be to connect with them.

I didn't. I spent much of my life desperate to please, which earned me that ironic nickname, "Guy Smiley." Over time, that changed; I grew freer to be myself, and to be alone. But sixteen months of pandemic lockdown curdled that freedom into avoidance. I expected to reemerge into the world salivating for connection with friends, old and new. Instead, social interactions felt like climbing a hill that got steeper as it drew closer. My instincts had shifted toward an unfamiliar shyness. That friendly remark to a stranger froze in my throat; spotting an acquaintance across the street, I took a step toward them, hesitated, and then turned back, ducking for cover inside a café.

I've spent years encouraging people to brave the social waters. But out of laziness and fear, I was failing to take my own advice. So, I recently decided to run a new experiment, which I called *encounter counting*. The premise was simple: For two days, anytime there was a chance to chat with someone, I would take it. I'd keep notes on each conversation, and at the end of it all I would tally how they had gone.

I chose a work trip to North Carolina, which included four flights, six meals, and two trips to the gym. The assignment seemed easy enough until I arrived in my seat for the five-hour flight from San Francisco to Washington, DC. Noise-canceling headphones are my typical armor on trips like this; leaving them in my backpack felt like going into battle wearing pajamas. The man who sat next to me looked kind, but probably wanted to keep to himself. I felt certain I'd be *that* person who tries to socialize on a plane, he'd cover his disinterest in the barest politeness, and the conversation would implode into three hundred minutes of awkward silence. I could feel the social sharks circling.

Someone walked by and complimented my seatmate's sweater, powder blue dotted with marigold. It was a great sweater, and I seconded the comment. "That's the last time I let my wife pick my clothes for me," he replied. "She's too stylish!" And we were off. He was a Muslim refugee from Sierra Leone who had found his way to the American South, gone to business school, and climbed his way through companies large and small. "I don't know if I'm the American dream or the American nightmare," he said. "Depends on the American," I replied. The conversation deepened quickly. We talked about parenting—his son, who has autism, changed his priorities forever—and about home. He told me about lowering his grandmother's body, by hand, into a plain hole in the ground, as is the tradition for funerals where he's from. That moment, he said, stamped any youthful arrogance out of him.

After an hour or so, a lull gave us both a chance to pull out our laptops, and we sat in 240 minutes of very unawkward silence. The plane landed, and we parted, exchanging only first names. The initial datapoint of my experiment was a rocket ship, launching me into a weekend of fun interactions. I sat at the bar of every restaurant, bringing a novel but eyeing neighbors for possible connections. At a café, someone was wearing the same unusual watch as me. At a French restaurant, the bartender and two diners debated which desserts were most universally loved.

Each time, I jumped, attempting eight conversations in all. Nervous about imposing, I left off-ramps for everyone. Determined to avoid creepiness, I quickly blurted out some mention of my wife to women. Most of these defenses proved unnecessary. At the end of the experiment, I scored five conversations as pleasant or very pleasant, one (my first seatmate) as extremely pleasant, and two as neutral. Some were perfunctory, but none were painful.

What shocked me was how much this shocked me. Over the last dozen years, my lab has done research with tens of thousands of people from the US and beyond. I *know* that the average human being is kind and open. I teach these principles. But deep inside myself, this knowledge had failed to register.

In a culture obsessed with measuring absolutely everything, "uncounting" can help restore peace of mind. But there are other parts of life that deserve more of our attention, including the positive moments we share with others. In these cases, keeping a tally can raise our awareness and help us savor social connection.

A weekend in Charlotte didn't reverse my pandemic shyness, and a single experiment probably won't change your life either. But now, when I hesitate to connect with someone, I realize it probably has nothing to do with the other person. Any of us can watch ourselves and our interactions more closely. If you're surprised at how warm people are, try to learn from that moment, so you're less surprised the next time.

The Startling Power of "Other Care"

After Watanabe spoke with his mother that first day, he peered back into his room—with the fresh, surreal perspective of an outsider. Its disarray and filth, "a condition that could never be shown to others," said everything about the mind of its resident. For Watanabe, his room was singular, coated in shame. But more than a million Japanese people also lived in their own reverse cocoons, slowly becoming more sealed in.

Exhuming his camera from the mess, he photographed himself and his room. It was an act of defiance. "I needed to turn the months of withdrawal upside down, sublimate them, and turn them into a counterattack," he tells me.

It took him nearly two years to adjust to the world outside, but in that time, Watanabe's art was reborn. Rather than running away from the *hikikomori* experience, he examined its contours. In one of his first art pieces after returning to the world, Watanabe built a tiny concrete home inside a gallery and walled himself inside. For seven days, he lived in a space no larger than the tatami mats Buddhist meditators use. At the end of the week, Watanabe chiseled himself out into the light, this time to camera flashes. The work, more personal than anything he would have dreamed of sharing before, finally found his audience.

Art, like a chemical reaction, can transmute one experience into another. In Watanabe's hands, isolation went from a personal trap to social commentary. He then began shining a light on others' pain, beginning with his mother. In a video project created six years after his isolation, the two sit across from each other, a tabletop clay model of their home between them. They demolish the structure with hammers in a few seconds, then spend hours gluing the pieces back together while discussing what their family has been through.

The meticulous repair invokes *kintsugi*, the Japanese tradition of honoring imperfections in pottery. It also reminds me of how things unfold for all of us at some point. We screw up; some part of our life shatters into a puzzle of our own making. All there is to do is slowly recombine whatever pieces we can find. That's what Watanabe and his mother have done. He lives on his own now, but she visits him often in his studio, twenty minutes away from the family home. His father doesn't know where it is.

For another exhibit, Watanabe invited *hikikomori* from around Japan to share photographs of their rooms. Dozens did, and Watanabe arranged them in a gallery behind a broken wall. To see the images, onlookers squinted through its cracks like voyeurs. They were confronted with a reality that defied stereotype. Not all recluses were hoarders, alcoholics, or gamers. Some kept their space tidy or were surrounded by religious objects. Each room had its own voice. Together they created a chorus, and a community for people defined by aloneness.

The opening drew large crowds of media and onlookers. Unexpectedly, some of the *hikikomori* who sent in photos arrived as well, emerging for the first time in months or years.

Watanabe has come full circle: from feeling oppressed by the art world to wielding his influence within it, using his creativity to give voice to vulnerable people. His work is now regularly commissioned by galleries and museums around the country. This transformation may have healed him personally, but that was never his intention. "I don't consider my creative activities self-care at all," he tells me. "I believe that helping and caring for others is the way to improve society. I believe that by improving society, I will be saved."

By focusing on social change, Watanabe unwittingly hits on the true origins of "self-care." In the nineteenth century, Peter Kropotkin, the Russian naturalist and prince (and later anarchist and prisoner), trekked through Siberia observing wildlife. In that inhospitable tundra, animals worked together: Wolves hunted in packs and horses gathered in defensive formations to repel them. Deer foraged for new pastures together, and birds huddled to stay warm. In his book *Mutual Aid*, Kropotkin argued that cooperation, not competition, was life's prime vector. To survive in hostile settings, animals had to tend to one another.

Society places marginalized people in their own forbidding landscapes. Urban "food deserts" lack proper nutrition; disabled people face inaccessible architecture and public spaces. Rather than waiting for the world to soften, many people form their own mutual aid communities. In the 1960s, the Black Panther Party launched "survival programs," providing medicine, healthy food, and yoga training to poor neighborhoods. These activities were simultaneously a form of care and of protest—asserting the humanity and worth of Black people against a culture that denied it.

In 1988's *A Burst of Light*, writer, professor, and activist Audre Lorde wrote, "Caring for myself is not self-indulgence, it is self-preservation, and that is an act of political warfare." You can now buy that sentiment on mugs and posters. Caring for ourselves, of course, is crucial. But by focusing people on the individual rather than the group, the self-care industrial complex loses the term's original meaning.

In the hands of Lorde and others, self-care is grounded in community and solidarity; it aligns with the nature of life as Kropotkin saw and what psychology and brain science teach us about humanity: There is no tidy separation between self and other. Our species is intertwined, such that helping others is a kindness to ourselves, and watching over ourselves supports others.

Mutual aid programs are an old tradition that lives on. Thousands sprouted during the pandemic. One neighborhood and Google spreadsheet at a time, people offered to shop for immunocompromised neighbors, pool funds for laid-off service workers, and stock pop-up food pantries. In the American West, suicide rates climbed among farmers. Communities responded by creating the "Coffee Break Project," in which people check on others who might be struggling. The program's tagline reads "Do you look after your neighbors as close as your crop or herd?"

Mutual aid could easily become a bigger part of our lives. Workplaces and schools could complement self-care activities with "other care": regular, structured chances to engage in kindness, such as monthly days of service in which employees and students volunteer together. Leaders—in companies, schools, teams, and towns—could build Ocean Villages wherever they go. But for many, that would mean reversing the ways they lead now, which too often reflect cynicism and cause it to spread.

Chapter 7

Building Cultures of Trust

On January 30, 2014, Bloomberg published an article bluntly titled "Why You Don't Want to Be Microsoft CEO." Despite enormous success, the company had botched early leads in e-readers, operating systems, and smartphones. In the year 2000, its market cap had been over $500 billion, more than a hundred times as large as Apple. By 2012, Microsoft had lost half its value, while Apple had ascended to a cap of $541 billion. The iPhone alone was bringing in more revenue than all of Microsoft's products combined.

A tech industry Goliath had been walloped between the eyes by leaner, hungrier companies. Microsoft's public failures stemmed from a more private problem. Its culture was drowning in mistrust, backstabbing, and obsession with short-term profit at the cost of long-term vision. The engineer and cartoonist Manu Cornet captured this in an "org chart," depicting the company's divisions in an armed standoff.

The problem was big, but not new or unique to Microsoft. *Organizational cynicism*—the feeling that one's community is full of greed and selfishness—has overtaken countless workplaces, destroying morale, well-being, and productivity. Many of these problems reflect an old misunderstanding of what allows any company, team, or school to thrive.

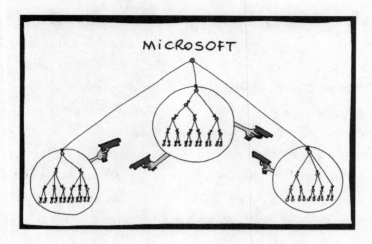

Homo Economicus Comes to Work

In its heyday, General Electric (GE) was nicknamed "Generous Electric," a nod to its corporate conscience and the huge amount of revenue it devoted to employee pay and benefits—ten times more than it gave to shareholders. GE leaders didn't consider profit sharing a luxury, but a necessity. As one executive put it, "The employee who can plan his economic future...is an employer's most productive asset." Workers who felt safe and secure would invest time, energy, and innovation back into their firms.

At least some of GE's largesse had been forced on it. In the hangover of the Great Depression, the US government imposed regulations to curtail corporate excess. Unions had ascended and were pushing for loyalty toward workers. But by 1981, politicians were again loosening the reins. Into this landscape stepped Jack Welch, GE's new CEO. Welch's vision would result in his being crowned "manager of the century" two decades later. What did he see? *Homo economicus.*

A century before Welch's time, the British economist John Neville Keynes complained that his field typecast humanity as creatures "whose activities are determined solely by the desire for wealth"—not human at all, but another species he named *homo economicus*. You probably wouldn't want *economicus* as a friend, colleague, or spouse. He is calculating and relentless in the pursuit of personal gain, and willing to toss any principle or relationship aside to get it. Thankfully, he's rarely seen in the wild. Economists themselves have documented many ways in which real people are kinder, more communal, and more principled than *economicus*.

From birth, *economicus* was meant as a caricature, not a flesh-and-blood species. But he will not go extinct. His first victims were economists themselves. When economics majors start college, they are socially similar to students in other fields. But research finds that by the end of college, they are less generous and more cynical. Simply learning about *economicus* makes them believe in fundamental selfishness—and then live it.

Many academics and leadership gurus also missed Keynes's joke. They heralded *economicus* as a genius, deciding that greed is a fast path to success. This fit perfectly with a pseudoscience that was popular in Keynes's time: "social Darwinism." In this line of thinking, philosophers and writers mutated Charles Darwin's theory of evolution to claim that human society is a war of one against all for survival. Darwin himself was not a social Darwinist, and the theory falls apart under any serious scrutiny. But it was catnip for the ultra-wealthy (and eugenicists, and the Nazi Party). People who hoovered up the nation's resources could use social Darwinism to justify themselves. Extreme inequality wasn't a moral failing, but a sign of some people's biological gifts. The tycoon John D. Rockefeller captured this enthusiasm, saying, "The growth of a large business is merely a survival of the fittest...It is merely the working-out of a law of nature and a law of God." The social Darwinist torch was passed across the century, from steel tycoons to hedge fund managers. In the film *Wall Street*, the amoral investor Gordon Gekko, played by Michael Douglas, updates Rockefeller for the 1980s, proclaiming, "Greed is right. Greed works. Greed clarifies, cuts through, and captures the essence of the evolutionary spirit."

Plenty of real-life Gekkos were inspired by this reasoning. It meant people and companies could maximize profit over any other value. If nature built us for endless competition, why should we deny it? Business schools drank deeply from this well, encouraging leaders to treat their people like packs of *economicus*. This meant unleashing them, red in tooth and claw, against everyone—including one another. According to one management professor, students learned that "companies must compete not only with their competitors but also with their suppliers, employees, and regulators."

Few leaders listened more closely than Jack Welch. Immediately after rising to power, he stripped away GE's generosity and motivated people through naked rivalry. His lieutenants described a "campaign against loyalty," where employees were thought of as liabilities, not assets. And, like costs, Welch cut them, conducting mass layoffs each year and replacing long-term employees with contractors who lacked job security and benefits. One of his favorite tactics was called "rank and yank." Each manager was forced to divide their teams into high, middle, and low performers. Standouts were rewarded; laggers let go. Welch viewed this strategy through a social Darwinist lens. The herd could run faster and further after being culled of its weaker members.

This philosophy crushed connections at work, but—for a while—increased profits. Generations of leaders extolled Welch and followed his lead. Fresh out of college, Steve Ballmer shared an office with Jeff Immelt, who would become Welch's successor at GE. Ballmer took over Microsoft in 2000 and brought the Welch playbook with him. He implemented rank and yank and imposed tight restrictions on employees, requiring layers of approval for even minimal decisions.

Meant to rein in workers, these policies slowed them down instead. The product manager Marc Turkel began a project at Microsoft in 2010, at the same time construction began on a block-wide, twelve-story building across the street. As Turkel negotiated across divisions, meetings compounded. Seeking favor from supervisors, their bosses, and their bosses' bosses, the team lost months. During yet another meeting, Turkel looked out the window and saw that the building across the street had been completed, while their project had no end in sight.

Under Ballmer, Microsoft also went to war against other tech giants. Many of their products were not allowed on iPhones. Instead of admitting that Apple had won the mobile wars, Microsoft made a last-ditch effort to compete by acquiring Nokia in 2013, resulting in a journey to nowhere that lasted years and cost the company billions.

Welch, Ballmer, and countless other CEOs today act as though their organizations are full of selfish, calculating free agents. Research finds that leaders' cynicism makes its way into both the carrots and the sticks they use as motivation. In cynical organizations, workers are rewarded for their individual performance even if they are awful coworkers—a "culture of genius" that erodes trust. Cynical managers also assume selfish workers will get away with whatever they can. To prevent this, they use preemptive strikes to monitor, threaten, and cajole labor out of people.

Factories, fulfillment centers, and cubicles have long been watched over by suspicious bosses. Eight of the US's ten largest private companies track individuals' productivity, often in real time, sometimes using methods that verge on the absurd. A hospice chaplain in Minneapolis reports that in 2020, her company started doling out "productivity points." Visiting a dying patient counted for one point; attending a funeral scored 1.75.

During the pandemic, millions of employees went remote, trying to get work done while managing early COVID chaos. Large companies could have responded by trusting them to meet goals in their own time. About 60 percent made the opposite choice, deploying a fleet of dystopian spyware to look over workers' shoulders. Employees were paid per minute of "active time," while their keyboards clacked, and cameras detected their faces. After bathroom breaks, one lawyer had to pose in front of her laptop at three angles, proving she had re-chained herself to the desk.

By comparing and monitoring, bosses put their workers into battle with one another while signaling mistrust in all of them. And as these practices became more popular, the corporate world's preexisting conditions skyrocketed alongside them. In 1965, US CEOs earned twenty-one times as much as their average employee. In 2020, they made more than 350 times as much. In a world full of *economicus*, this makes sense. In ours, what does it cost?

The Heavy Price of Mistrust

At Microsoft, it cost quite a lot. Under Ballmer, morale collapsed, and so did innovation. One major culprit: rank and yank. Every six months, managers drew the blinds on conference rooms and arranged Post-it notes on a whiteboard to decide their employees' fates. No matter how talented their teams, most had to choose some people as poor performers, destined for the chopping block.

The effects cascaded throughout the company. Talented engineers did their best to stay away from one another, preferring to be on top of a mediocre team rather than in the middle of an exceptional one. In a zero-sum world, surviving didn't mean excelling. Running from a tiger, employees just had to be faster than their slowest colleague—or to trip them. As one engineer explained, "People responsible for features will openly sabotage other people's efforts. One of the most valuable things I learned was to give the appearance of being courteous while withholding just enough information from colleagues to ensure they didn't get ahead of me on the rankings." Employees also circulated rumors about one another, which the product manager, Marc Turkel, called "management by character assassination." In East Germany, spies could be at the store, on the sidewalk, or in your home. At Microsoft, they hung out by the watercooler.

Collaborations fell apart as employees bickered over who would get credit. People stayed in their comfort zones to avoid failure. To maximize shareholder value, teams chased short-term profit instead of new trends that could pay off in years or decades. Microsoft grew brittle and bureaucratic.

Employees of cynical organizations are more burned out and less satisfied in their jobs. They butt heads more often and keep knowledge to themselves. Cynicism quickly trickles down, from leaders to their employees. Edelman's 2022 "Trust Barometer" found that when employees felt trusted by their boss, they trusted right back 90 percent of the time. When they felt mistrusted, less than half had faith in their own managers, and barely a quarter in their CEOs. And like cynical lives, cynical jobs tend to be shorter—as workers look for the exits.

You may suspect that all this pain is the cost of winning. The steel titan and social Darwinist Andrew Carnegie thought so. Endless struggle "may be sometimes hard for the individual," he regretted, but "it is best for the race, because it ensures the survival of the fittest." That would be sensible enough, except that cynical organizations turn out to be remarkably unfit.

In a rush to naturalize greed, social Darwinists ignore what Peter Kropotkin and other scientists discovered: Animals thrive by working together, not against one another. Hyper-social creatures, like human beings, take this to another level. Communities form their own cultures and compete against one another like clashing superorganisms. The more intense the struggle between groups, the more individuals must team up to come out on top. During wartime, nations and tribes hate the other side but swell with patriotism for their own. Dozens of studies from over forty countries have found that war intensifies generosity within groups, for instance, when soldiers and neighbors risk their lives for one another.

Conflict doesn't have to be violent to bring each side together. In sports, work, and life, the more a group cooperates, the better it fares against the competition. Bill Bradley, the NBA forward turned US senator, put it well, reflecting that "the success of the group assures the success of the individual, but not the other way around." Rank and yank, surveillance, and micromanaging destroy these advantages. Organizational cynicism also costs gobs of money. Colleagues and firms who trust one another trade often and build long, mutually beneficial partnerships. Cynicism replaces this with doubt and friction. Informal agreements give way to baroque contracts; lawyers are brought in and paid handsomely to protect everyone from everyone else. The resulting fees, known as "transaction costs," can quickly scale to tens of millions of dollars.

By creating a world of bottom-line-obsessed *economicus*, cynical organizations lose their humanity—and the bottom line. Kicking, clawing, and gouging your colleagues, you might be able to make it to the top of a community like this. Once you do, you might find that the firm, team, or community next to you has been cooperating all along. The few who survive a cynical Lake Town, bloodied and haggard, can seldom match an Ocean Village working as one.

Growing Cynics

Few of us are born for backstabbing and office politics. Early Microsoft employees remembered a different company: engineers in Hawaiian shirts; an atmosphere of cheerful nerdiness. But by treating workers as though they were selfish and untrustworthy, Microsoft—and GE, and countless other firms—coaxed the worst out of them.

In Boston, firefighters took more sick days after being accused of fleecing their department. At Wells Fargo in the 2010s, managers forced employees to race one another toward unreasonable sales targets. They responded by opening hundreds of thousands of fraudulent credit card and bank accounts, resulting in nearly $200 million in fines for the company. One employee described feeling forced to trick an elderly woman into opening an account she didn't need as "the lowest point of my life." In *The HP Way*, David Packard writes about his time at GE, when the company locked up equipment to prevent theft. "Faced with this obvious display of distrust," Packard writes, "many employees set out to prove it justified, walking off with tools and parts whenever they could."

Now that distrust has gone digital, so has retaliation. Cottage industries have popped up to help workers fool their company's spyware. "Mouse jigglers," like miniature Roombas, move a person's computer mouse at random intervals, creating the appearance of work. One popular jiggler amassed thousands of five-star reviews on Amazon, including this gem: "If your boss is a micro-managing worthless idiot who doesn't realize that presence does not equate to productivity, this is the device. If you are one of those bosses reading this review...nobody likes you." As the website dutifully informs me, "389 people found this [review] helpful."

Economicus started out as a joke. Bad leadership breathes him into life, leaving everyone worse off. This pattern plagues modern work, but it doesn't stop—or start—there. The first organization most people join isn't a company; it's their school. And many classrooms operate with a level of cynicism that would make Jack Welch blush.

In her early forties, LaJuan White could no longer live on the block where she had grown up. The Brooklyn middle school principal watched as rent prices crept, then soared, past what most educators could afford. In 2015, she'd seen enough and requested a transfer to Syracuse. The district assigned her to an elementary school, but three days later the superintendent redirected her to Lincoln Middle School.

One look at her new assignment could strike fear into even a veteran educator. Lincoln Middle had the fifth-highest suspension rate in the state. It ranked among New York's "persistently dangerous" schools, defined by having more than three serious violent incidents per one hundred students each year. In just six years, Lincoln had chewed up and spit out four principals. From the outside, the brick and teal building seemed like any other school, but what horrors awaited White inside? She didn't know.

A week after starting the job, she still didn't. The kids could be mean to one another. Some were chronically absent. But these were preteens, not menaces to society. A quarter had special learning needs and many had arrived in the US as refugees. What White did notice was a "culture of punishment." Teachers expected kids to act out and held tight to their best counterstrike: "Violent or Disruptive Incident Reporting," or VADIR (pronounced "vader," like the *Star Wars* villain).

Teachers used this system to document assaults, drug dealing, and weapons possession. Other VADIR categories, such as bullying, were fuzzier. A rude comment might be just rude, or it might be reportable, depending on a teacher's judgment—which depended too often on the student they judged. Black and white students were suspended at comparable rates for clear offenses (bringing a knife to school), but students of color were punished more for subjective infractions, like being "disrespectful." Reporting an outburst was the quickest way to stop dealing with it. So, many teachers did, and thus developed a keen eye for the students' dark sides.

Syracuse is starkly poorer than neighboring towns—many Lincoln teachers had chosen it because of this, hoping to combat educational inequality. These were idealists, but the punishment culture brought out their most cynical side. "When adults are in their feelings," White told me, "they end up describing the worst possible version of that child." This mistrust spread to kids, and why wouldn't it? When students are punished excessively—or even see their friends mistreated—their faith in school collapses, and they act out even more. As one scientist writes, kids under a punishment culture "engage in more defiant behaviors to reassert their freedoms and express their cynicism toward institutions." At Lincoln, students became exactly who teachers feared.

Leading *Homo Collaboratus*

The cartoon of Microsoft's standoff bothered Satya Nadella. Nadella had stepped into the role of CEO just five days after Bloomberg's article said no one should want the job. But what upset Nadella more than the cartoon itself, he later wrote, "was that our own people just accepted it." As a leader, he set out to restructure the company's culture according to a different vision. Ballmer had managed as though his people were *homo economicus*.

Nadella assumed the opposite: that Microsoft was full of *homo collaboratus*, a species who wanted to create together.

Along with his chief people officer, Kathleen Hogan, Nadella created an environment for *collaboratus*. Rank and yank had been pulled the year before, and the new leaders introduced a more integrative review system. Now employees were evaluated not just on individual performance, but also on how they supported one another. This wasn't simply a nice way to reward workers; it was wise. When people's outcomes are connected, they are more likely to join forces. Researchers call this "task interdependence." Like a photonegative of zero-sum thinking, task interdependence increases trust between colleagues and brings them closer. It also makes work more efficient—as people share knowledge freely and work in sync toward common goals.

Nadella even treated other firms like *collaboratus*. In a move that would have been unthinkable years before, he strolled onstage at an industry keynote, reached into his pocket, and produced an iPhone—equipped for the first time with Office, Outlook, and other Microsoft products. Nadella was, in essence, conceding the mobile technology race, but allowing both companies to win by giving customers what they wanted. "Partnering is too often seen as a zero-sum game," he wrote. Inside and out of Microsoft, Nadella sought opportunities to grow the pie and tap into collaborative instincts.

In the new Microsoft, tight management gave way to massive "hackathons," team coding sprints that encourage a free-for-all of new ideas. Leadership gave people more space—and listened to them more closely. During the early pandemic, Hogan launched a series of surveys to understand employees' experiences. The message that came back was clear: People were fighting through a hurricane of unpredictable childcare, sickness, and worry. They needed flexibility and support. In response, Microsoft announced that employees could work from home for the long term, expanded mental health benefits, and added twelve weeks of parental leave. At the same time, Hogan invested in leadership training to help executives remain as supportive and connected as possible. It worked: In 2020, more than 90 percent of Microsoft employees trusted in their managers, and Hogan was dubbed 2021 HR Executive of the Year.

Microsoft's leadership received trust by giving it first—but that's not all they gained. Those freewheeling hackathons got employees thinking in more agile terms. New ideas took hold. Engineers explored cloud computing and artificial intelligence, and the company made enormous investments in OpenAI, creators of ChatGPT. Microsoft's market cap soared by nearly tenfold in just a few years. Instead of a competitive advantage, Nadella discovered a cooperative advantage that emerges when leaders put faith in their people.

Lincoln Middle could scarcely be more different than Microsoft's sleek Washington campus. But the philosophy that LaJuan White pursued there looks a lot like Nadella's. Punishment culture assumes the worst about kids. White began with questions instead. Why were students acting out? What were they going through?

To find out, she began a "porch tour," visiting many students' homes. Families were confused at first, then grateful for the attention to their kids. By connecting family and school, White cultivated a more personal type of authority. "I was on a first-name basis with many parents," she recalls. "If a child got saucy, I could tell her, 'Don't make me call Paul!'"

She also saw firsthand what kids were dealing with. In one "troublemaker's" home, a broken window had been taped over with thin plastic, which waved rhythmically in the frigid breeze. Vermin scuttled across the floor. It was clear that the child ran the home, and school was their only place to be a kid. "This completely changes your outlook," White tells me, "from 'What did you do wrong?' to 'What do you need? Are you hungry? Do you need time to shake something off?' The ones who seek attention in the most unconventional ways are the ones who need love the most."

Next, White sought out to reform Lincoln's punishment culture. For guidance, she turned to the philosophy of restorative justice. When someone is wronged, punitive justice involves finding the culprit and making them pay. A restorative approach instead asks who was harmed, and how their pain can be resolved. "If you've made a mistake," White says, "and I suspend you, where is the learning from that? Instead, we want to know how both parties can heal from a violation."

Teachers weren't having it. If they couldn't VADIR a child, their classrooms would fall into disarray. They felt micromanaged, disempowered, and unsafe. White was cussed out "more than once," and each day before getting in her car she'd check the tires, worried a disgruntled adult might have slashed them. But she didn't back down. White set up a dialogue with teachers, listening to their concerns, and inviting them to cocreate a new plan with her.

The result was a recipe for restorative discipline, which Lincoln named the "classroom hierarchy." If a student acted out, teachers redirected them. If that didn't help, they pulled the child aside for a one-on-one conversation to learn what caused their outburst. Students then spent ten minutes in self-reflection. If and only if those strategies failed, teachers could send the student out of class or school. The new system demanded more of adults. Jen Harris, an educational coach at Lincoln, said, "I have to resolve [disciplinary issues] on my own. No teacher can go to the administrators and say 'Do something with this kid.' Because it's like, 'What have you done restoratively? How have you built a relationship?'"

Acting differently toward kids, teachers began thinking differently. A misbehaving child was less a danger to be removed and more a partner to work alongside. Research finds that when teachers show faith in their students—especially during struggles and conflicts—students reciprocate trust and are less likely to fall through the cracks. That's exactly what happened at Lincoln. White watched, sometimes on the verge of tears, as "rough" kids hugged in the hallways and bullies made amends with their victims. Suspension rates dropped by more than 15 percent in White's first year. The school quickly shed its "persistently dangerous" label and hasn't gotten it back since.

Learning Anti-Cynical Leadership

Leaders set the preexisting conditions in which people live, learn, and work. A punishment culture shows kids a version of themselves beyond hope and help. Jack Welch–style management shows employees that they are *economicus*, and shouldn't bother trying to be anything else. But we can do better. White committed to a simple mantra: Treat kids like the people we want them to be, not the ones we fear. Nadella unwound cynicism at Microsoft and gained a cooperative edge.

You don't need to run a school or a trillion-dollar company to follow their lead. Emile designed his Peace and Conflict Neuroscience Lab for *homo collaboratus*—for instance, emphasizing that scientists should share credit rather than trying to outshine others. I wish I could say the same. In 2012, Stanford gave me a nice office and some funding to start my research lab—a dream come true after nearly a decade of training. But stepping in, I realized how ill-equipped I was for the job. I felt like a college athlete who is drafted into the pros—to be a coach. It was the same game but required a totally new skill set. Assistant professor jobs are also temporary; you have a few years to do the best science you can before applying for tenure. Then, the most accomplished people in your field decide whether or not you should be fired.

This process did not play well with my anxiety, and the result hurt the people around me. I felt intense pressure to produce, and for the first time, couldn't do it alone. My future depended on people I had hired weeks earlier. I reacted by pushing the young scientists in my lab hard, checking in on them frequently, and—more than I realized—voicing disappointment when they didn't meet my unrealistic standards. It was textbook cynical leadership, and quickly backfired. About a year into my life as a professor, someone in the lab asked to meet with me urgently. In tears, she said the stress of her job had grown unhealthy, and unless things changed, she would leave.

Every detail of that conversation lives on in my mind; nearly a dozen years later, the memory flushes my face with shame. In a bitter irony, an empathy scientist had created a toxic culture. I'm also grateful for that lab member and the bravery she displayed in challenging a new boss. Her feedback woke me up like a bucket of ice water and I realized I needed to change my management style. I committed to meeting my people where they were and trusting them to meet goals in their own way.

Though I hadn't heard the term yet, underbearing attentiveness—Bill Bruneau's parenting style—became my north star as a leader. At the beginning of each meeting, rather than asking students what they had done, I asked what they needed from me. I did my best to follow their lead, managing closely when they asked, and letting them explore when they preferred. We created a "lab manual" codifying our values, with cooperation at the center. The entire lab took part in reviewing the manual, giving them ownership of the process, and creating a clearer understanding of what we all expect from one another.

These changes calmed my nerves and improved moods across the lab. I decided that even if this new style cost me tenure, that would be better than achieving it by leading poorly. But the opposite occurred. The leaps of faith I took on my students were paid back in the form of innovative, diligent science. In a looser culture, lab members worked together more, generating ideas none of them would have come up with alone. The cooperative advantage arrived in our little community and stayed.

More recently, I've tried to impart these lessons to others. At the software company SAP, I taught rising managers on four continents about the corrosive effects of cynicism and how they could use asset-framing and leaps of faith instead. When one of their colleagues struggled, they might rethink that person's tasks to bring their strengths to the surface. When managers gave someone a new level of responsibility, they could "trust loudly," explicitly calling out their faith in that person. Many of these leaders were new. Like me during my first year as a professor, some felt safe only when micromanaging their teams. But one by one, they unlearned these habits and led with trust. In response, their teams became more productive and their own employee ratings climbed about twice as fast as those of managers who didn't take part in the program.

Any leader can learn anti-cynicism. Soon, the ones who don't might find themselves left behind. During the pandemic's first two years, millions of people left their jobs in a "Great Resignation," while countless more "quiet quit" by abandoning as many of their duties as possible. Leaders have been angered and confused by these trends, but they shouldn't be. The campaign against loyalty started at the top decades ago, with suspicious, extractive management. The Great Resignation is simply a long-delayed response.

Rebuilding cultures of trust will require structural changes, like reducing workplace inequality and restoring job security. But this must be accompanied by a psychological overhaul, with people in power putting more faith in those who have less. Employees, for their part, are starting to demand more. Until recently, union membership was on the decline, dropping in the US from over 20 percent in 1980 to barely 10 percent in 2021. But that pendulum appears to be swinging in the other direction. Employees of companies such as Amazon and Starbucks have begun high-profile unionizing campaigns. In 2023, writers and actors in the entertainment industry went on strike, as did members of the United Auto Workers. Hundreds of thousands of Americans walked off the job, a rising tide that is also gaining popularity. A 2023 poll found that more than two-thirds of Americans support unions, up from less than half in 2009.

During the pandemic, workers realized their power. If leaders won't reverse their cynical practices, people might vote with their feet, for organizations that trust, value, and believe in them.

Chapter 8

The Fault in Our Fault Lines

In 1983, the USSR nearly started World War III out of fear that their rival already had. Operation RYaN had launched two years earlier (*Raketno Yadernoe Napadenie* is Russian for "nuclear missile attack"). It was the largest Soviet intelligence operation of the Cold War, built entirely on false premises.

Operation RYaN was the brainchild of KGB director Yuri Andropov. Andropov had been the Soviet ambassador to Hungary in 1956, when Hungarians joined together in a national uprising. After assuring Hungarian leaders he would not invade the country, Andropov did just that, leading a brutal suppression of the revolt. Soviet tanks fired into civilian buildings, killing thousands. Andropov became known as "the Butcher of Budapest," a figure of terror in Hungarian history. The experience also left scars on Andropov, who watched as soldiers were publicly executed by Hungarian resistance forces.

Having witnessed how quickly Soviet power could be threatened, he spent the rest of his life paranoid about its collapse. In the 1980s, he turned that obsession toward the US, sure it was planning a nuclear strike. He commanded dozens of officers to look through hundreds of potential clues. Were American troops amassing in unusual places? Were Pentagon parking lots full at night? Were blood banks collecting more donations than usual? Agents were urged to "report alarming information, even if they themselves were skeptical of it."

The result was a mess of false leads that—when squinted at from the right angle—painted a picture of imminent war. The KGB began thinking that the only way to not be blindsided by their enemy was to blindside the Americans first. If a double agent hadn't alerted the West about Project RYaN, this preemptive strike could have been humanity's last.

Andropov began with the assumption that America wanted war, and manufactured proof wherever he could. Similar fears have recently captured many Americans, except that their boogeyman is not another world superpower, but fellow citizens. The Oath Keepers are a right-wing militia who want to take power from the government. They are terrifying, especially considering that many work in government as soldiers and police officers. But their meetings and message boards make clear they are also *terrified*: certain that the forces of a shadowy deep state will come for them. In June 2020, protests erupted around the country following the police murder of George Floyd. The Oath Keepers' leader, Stewart Rhodes, urged his people to stay on high alert. "Let's not fuck around," he announced. "We've descended into civil war."

Rhodes's fear doesn't excuse his crimes. In 2023, he was sentenced to eighteen years in prison for his actions during the January 6 insurrection. Most Americans are not as violent or as paranoid as the Oath Keepers, but many of us share their conviction that war is coming. In 2022, a poll found that 69 percent of both Democrats and Republicans believed that national rule of law is under direct threat. Conflict is rising in the US and beyond, for lots of reasons, including *tribal cynicism*: the belief that people on the other side are stupid, evil, or both.

The Devil We Don't Know

Imagine the average person whose political beliefs are the opposite of yours. What do they look like? Where do they live? What do they do for work? For fun? What do you think their views are on immigration, abortion, and gun control? Would they support violence to advance those views? What would this person think of you?

During the same era in which Americans lost trust in one another, they grew contempt for people they disagree with. In 1980, US Republicans and Democrats felt lots of warmth toward their own party (from now on, we'll call them a person's "insiders") and neutral about the other (we'll call them "rivals"). By 2020, each party disliked the other side more than they liked their own.

People fear and loathe rivals—and increasingly avoid them. In the 1970s, the US had blue and red states like it does now, but many counties within those states were indigo or amethyst. Since then, Americans have "sorted," moving away from rivals so much that counties are as politically segregated as they were during the Civil War.

As people interact less across political lines, we lose real-world knowledge about rivals. That doesn't mean we stop thinking about them. The information vacuum fills with media and our imagination. As we've seen, both of these forces are dominated by negativity bias, and tip us toward cynicism.

Let's return to the "average" political rival you just imagined. Across dozens of studies, Americans have been asked to answer questions like you did; they answer incorrectly in virtually every way scientists can measure. Even outside of politics, we guess wrong about one another's lives. Democrats think 44 percent of Republicans earn more than $250,000 a year. In fact, only 2 percent do. Republicans think 43 percent of Democrats are part of a labor union, but just 10 percent really are. Republican cat people guess that Democrats favor dogs, and Republican dog-lovers think Democrats must like cats.

Both Democrats and Republicans also imagine rivals are more *extreme* than they really are, a pattern researchers call "false polarization." On issues like immigration and abortion, people guess the average rival is more extreme than 80 percent of actual people in the other party. Asked about the middle, we conjure up the fringe. The more specific the questions, the more wrong we become. Democrats think 35 percent of Republicans would agree that "Americans have a responsibility to learn from our past and fix our mistakes." The real figure is 93 percent. Republicans think 40 percent of Democrats believe that "the Constitution should be preserved and respected." In fact, 80 percent do.

False polarization mixes negativity bias with ideology. If I believe in something, my enemy must believe the opposite. These cynical misperceptions obscure an entire landscape of national consensus. In 2019, Emile and his collaborators asked Americans how much immigration should be restricted on a 0–100 scale, where 0 meant all borders should be open and 100 meant they should all be closed. They also asked them to guess how the average political outsider would answer that same question. They discovered two landscapes. One, made up of real people, looked like a hill with two peaks: Democrats wanted more openness, Republicans less, but there was lots of overlap in the middle. The second, made up of our perceptions, looked like two separate hills, each populated with extreme, unconnected opinions.

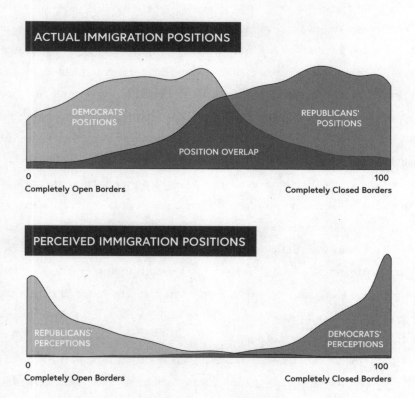

That image says a lot about the moment we're in. There's plenty of American agreement, and not just on immigration. A 2021 survey of more than eighty thousand people identified nearly 150 issues on which Republicans and Democrats agreed. Several were endorsed by more than two-thirds of *both* parties. These included overturning Citizens United so that companies could not fund political campaigns, giving immigrants who arrived in the US as children a path to citizenship, and tax incentives to promote clean energy. And yet, in our imagination, shared values have eroded into tiny islands, barely visible above the waves.

We don't know who's on the other side, what they believe, or how peaceable they are. Recently, scientists in twenty-six countries asked people how they felt about political rivals, and how they thought rivals felt about *them*. Both conservatives and liberals disliked the other side. But in nearly every country, people vastly overestimated how much the other side disliked them. Americans also imagine rivals are gearing up for violence. In 2020, researchers asked fifteen hundred people whether they would support violence to advance their side's cause. Five percent of Democrats and 8 percent of Republicans said they would: a small but frightening minority. But each group *thought* that more than 30 percent of the other side supported violence, dreaming up rivals four times more bloodthirsty than the real thing.

In the year 410, Rome was home to eight hundred thousand people. Its walls had fended off attackers for more than eight centuries. Then the Visigoths, a Germanic tribe long mistreated by the Roman Empire, laid siege to the city with outstanding stamina and breathtaking cruelty. Inside, citizens panicked and considered trying ancient sacrifice practices in desperate hope the gods would intervene. I imagine they must have seen the Visigoths less like human enemies and more like a relentless force of nature tearing down everything precious.

Many Americans now feel much the same way, as though the walls are being breached. Of course, there *are* many real political dangers in the US and beyond. But one surefire way to worsen them is to assume our rivals are Visigoths while they conclude the same about us.

War No One Wants

About twenty years ago, scientists brought pairs of strangers together and asked one to tap the other's hand. The second person's job was to tap back just as hard. The first person then matched the second person's tap, and so on. If they were accurate, the force of their blows should have stayed constant. Instead, each one *felt* like they'd been hit harder than they really had. Trying to reciprocate, they escalated. On average, each tap was 40 percent stronger than the last. Within a few rounds, people were hitting one another twice as hard as when they'd started.

In a polarized world, negativity bias skews political thinking and preemptive strikes shape our actions. Researchers find that people who imagine hateful rivals are also more likely to agree that their party should "do everything they can to hurt [the other party], even if it is at the short-term expense of the country." The same is true of violence. Individuals who thought—incorrectly—that the other side was spoiling for war became more supportive of violence themselves three months later.

The vast majority of Americans prefer peace. But when we imagine the other side is out for blood, we start mounting our defenses. Each side looks for evidence to confirm their fear, like Operation RYaNs in parallel. And there's no shortage of bad intelligence. Stewart Rhodes fed it to the Oath Keepers. In 2022, the author Malcolm Nance published *They Want to Kill Americans*, a book about conservative militias. We should all worry about domestic terrorist groups, but Nance generalizes, pre-disappointment style, to the entire right half of the US. "The average Trump supporter believes he's supposed to be preparing for civil war," he said in one interview.

Rhodes and Nance are wrong, but their fever dreams match our cynicism, so they gain influence anyway. Both are what journalist Amanda

Ripley calls "conflict entrepreneurs": people who inflame social division for their own ends. Cable news executives and partisan influencers make a living by depicting political rivals as modern-day Visigoths.

A favorite tactic among conflict entrepreneurs is known as "nut picking": showcasing extreme rivals as if they represent an entire group. Malcolm Nance wants us to see the average Trump supporter as Stewart Rhodes, and Rhodes wants us to see the average Democrat as an "antifa" militia member.

Online, any of us can join the culture war, retaliating against enemies both real and imagined. Just like professional conflict entrepreneurs, we're rewarded for doing so. Beginning in 2017, researchers analyzed more than twelve million tweets about politically charged issues. Posts including conflict words—such as "fight," "war," and "punish"—went viral more than neutral ones. Later, these same scientists messaged thousands of people after they tweeted and asked how they were feeling. Even people who felt relatively calm pretended to be outraged in their posts.

On some platforms, beautiful people use filters to appear more beautiful. On Twitter, extreme people filter themselves to appear more extreme. In each case, the rest of us end up with a skewed, cynical view of who's out there.

Most Americans don't hate one another. We do, however, hate how divided we've become. In my lab, we find that more than 80 percent of Republicans and Democrats say polarization is a major problem for the nation and would prefer more cooperation across parties. And yet many of us feel that we must flame, attack, and derogate rivals in self-defense. Trying to match their aggression, we escalate, taps quickly turning into slaps and punches.

This snuffs out people's sense of possibility. Our side might win or lose, but there's a growing sense that *all* is lost, and our entire national project has failed. This is especially true for young voters. In a 2021 poll, only 7 percent of eighteen- to twenty-nine-year-old Americans reported that their country had a "healthy democracy." Nearly twice as many, 13 percent, claimed it was a "failed democracy." In Israel and Cyprus—two countries under long-standing conflict—researchers asked over a hundred thousand people whether peace could ever prevail. The younger someone was, the less hope they expressed.

Young Americans grew up in the shadow of 9/11. In elementary school, their teachers conducted active shooter drills. By high school they learned that the natural world would wither before they reach old age. Young Israelis and Cypriots have never known their countries at peace. Nihilism is a perfectly understandable reaction to these circumstances, but it forecloses on the possibility of anything else.

Quickly, this attitude takes on the hallmarks of cynicism. Political hopelessness feels smart, and the idea that nations *could* find harmony starts to seem simpleminded and dangerous. Ironically, these cynical views about rivals are themselves clearly naive. But as political hopelessness charges on, it helps the most duplicitous political elites. So long as we think productive exchange is impossible, they don't have to work toward it. So long as we fight over partisan identities, the struggles that most of us share—such as rising inequality—remain under the radar.

Unwinding Misperceptions

Culture wars in the US and beyond feel eternal, but in the past things have been much better, and also much worse. Hatred has risen but could fall just as quickly. The vast majority of Americans want this, but fewer of us hope for it. If that changed, we might do more to bring it about.

In our personal lives, better data can make us less wrong and more inclined toward hope. Could that same strategy work in the face of deep conflict?

Andrés Casas didn't think so but ended up trying anyway. Casas grew up in Bogotá, Colombia, privileged but just a few miles from poverty and brutality. A Communist rebel army, FARC, had been locked in a struggle with the Colombian government since the 1960s. Kidnappings, rape, and torture were rampant. Across five decades of violence, more than two hundred thousand Colombians lost their lives and more than five million were displaced. FARC-held areas were deprived of government support, leaving bystanders without basic services and driving many into the drug trade.

As a teenager, Casas found solace in the hard-core punk rock scene. Beneath their pummeling sounds, groups such as Bad Brains and Youth of Today preached an egalitarian, Buddhist way of life. Under their influence, Casas dedicated himself to social change through science. He earned degrees in philosophy and political science while dabbling in several other fields. After years at the desk, his ideas felt ready for road testing.

In 2013, a university invited Casas to do research in Antioquia in the country's northwest. Two decades earlier, the region's capital, Medellín, had been the world's most dangerous city, overrun by drugs and open warfare between FARC guerrillas and government paramilitaries. Civilians were caught in the crossfire, surrounded by unthinkable, near-constant violence. Yet by the time of Casas's visit, the murder rate had plummeted in what became known as "the Medellín Miracle." He wanted to understand how the miracle occurred, and how it might be replicated elsewhere.

Casas created a pop-up laboratory, using surveys and experiments like the trust game to assess how Antioquians thought about one another. It immediately became clear how inadequate these tools were. Cartel members, many of whom had been part of the drug trade since childhood, lived by utterly different social codes than other people. Soldiers and citizens alike were drenched in collective trauma, which produced what Casas calls "mind freeze." Past violence had shocked the possibility out of people, making peace feel impossible. As one former FARC combatant described, "War is war, it never ends. The same as always, we're waiting for the roof above our heads to fall in." Casas's academic notions suddenly seemed laughably irrelevant. "I felt so limited in the face of real suffering," he remembers.

And yet under all that destruction, new possibilities were emerging. In 2010, Juan Manuel Santos won the presidency of Colombia. His predecessor had taken a hawkish stance against the FARC, but Santos thought differently. In an early speech, he offered that the "door to dialogue" with the guerrillas was not "closed with lock and key"—a small but powerful leap of faith.

Soon, FARC and government leaders entered exploratory talks, and by 2012, a full-scale peace negotiation was underway. The process was tenuous and halting, interrupted by spurts of violence and many crises. But by 2016, the two sides had struck a peace deal. The government offered to establish FARC as a legal political party. Combatants would be prosecuted by a special court but would be given more-lenient sentences if they confessed. FARC promised to disarm and pay reparations to victims.

An end to the violence was in sight. But first, the nation would have to support these measures in a national referendum. It did not. Colombians voted down the peace agreement by a razor-thin margin of 50.2 to 49.8 percent. After a long collective nightmare, half the population voted not to wake up. Casas, along with millions of others, was confounded. He decided that to help his country he needed to understand the mind, and began a master's program in psychology at the University of Pennsylvania. There, he happened into a conversation with a local professor: Emile Bruneau.

The two scheduled a meeting to talk about research. Planned for fifteen minutes, their conversation lasted more than ninety. When Casas's home country came up, Emile asked, "Why did Colombians vote against peace?" Casas had no simple explanation but offered one intriguing fact. Colombians in rural areas, closest to the conflict, voted for peace in larger numbers. People insulated from the violence were most likely to vote no. In those data, Emile sensed an opportunity, and he made Casas an offer: "Come work in my lab, but the commitment will be first, to get a proper understanding of what happened in Colombia; and second, to do something about it."

Soon, Emile and Casas were on a plane to Colombia, planning to follow Emile's approach to peace science: diagnose, then treat. In the months leading up to the peace referendum, the "No" campaign, advocating against the peace deal, ran a media blitz to sow fear among Colombians. The message, repeated on every television, phone, and tablet, was that all FARC members were unrepentant murderers who could never rejoin society. Ads included cartoons that depicted the FARC as rabid monsters preying on the average Colombian.

Among the "No" campaign's target audience was Casas's mother. A dedicated public servant, one of her duties was providing aid to victims displaced and tortured by the FARC. Through a parade of brutal stories, she came to despise the FARC and was uninterested in her son's dreamy notions about peace. The family had a rule: "Never talk about politics. Ever."

The "No" campaign was wrong in at least one clear way. Ex-FARC members *were* reintegrating peacefully into communities, but in rural camps, far away from cities. Most Colombians had never met a FARC member. One way to correct their misperceptions would be to introduce ex-FARC members to the public. Research has found, time and time again, that when people interact with outsiders one-on-one, some of their prejudice melts away. In Colombia, that type of contact could never happen at scale. But if the anti-peace campaign used media to feed fear, perhaps a different kind of media could feed hope instead.

Andrés Casas's brother, Juan, is a filmmaker, and he agreed to join the effort. Emile and the Casas brothers traveled to a demobilization camp to interview ex-FARC members. As they rumbled through the hills of Antioquia on a bus, Emile's reflections were caught on camera: "I have a feeling that what [ex-FARC members] have in their heads does not match up with what Colombians think is in their heads, and that's the whole premise of this experiment right now. I don't know how right I'm going to be, so there's some anxiety about that." He had other reasons to worry. UPenn couldn't directly pay interviewees, so Emile was funding the project on a personal credit card. "I hope the police don't come knocking at my door," he told a colleague.

The team's first test came quickly. The young Colombians in Juan's camera crew had been inundated with the same media as the rest of the nation. They brought pre-disappointment to the countryside: expecting FARC members to be subhuman terrorists and drug dealers. The interviews shattered these ideas. Ex-FARC leaders, combatants, and nurses shared their trauma, regrets, and hope. Most were "campesino" peasants living in poverty; many had watched as children while their grandparents, parents, and siblings were killed by government paramilitaries. The FARC were villains in the nation's story, but victims in their own. Most of all, nearly every person Emile and Casas interviewed wanted peace.

The camera crew were also the project's first audience; Emile and the Casas brothers asked what they thought. "I'm really confused and doubting all the stuff I've learned," said a young woman, "because if you get to know these people and their version of this war…if we were in that position, we absolutely [would] have done the same to fight for our families." Another, her voice cracking with emotion, replied, "I don't know how to describe what I'm feeling, but I'm thankful I can be here."

The team interviewed neighbors of the camp as well. Unlike most Colombians, they coexisted with ex-FARC members—and knew that such coexistence could be peaceful. Juan Casas edited the footage into five-minute videos that showcased the humanity of ex-FARC members and the reintegration occurring in Antioquia. The team then conducted an experiment. They showed these clips to hundreds of Colombians, and showed hundreds of other Colombians videos unrelated to the conflict. Their films were a peace-enhancing medication, tested against placebos. After people watched one or the other, Casas, Emile, and their collaborators surveyed viewers about their attitudes toward the FARC, and toward peace.

The results were dramatic. Colombians who watched the videos of FARC members came away skeptical of the prevailing, hopeless narrative.

They were more likely to believe that ex-FARC members wanted peace than did the people who watched "placebo" videos. They were also more likely to support reintegration. This yearning for peace remained with viewers even three months after watching. For Andrés Casas, the strongest proof came from his mother. Weeks before her death, he showed her what he was working on. After watching the video, she told him, "Now I get it. I see why you're doing this." For years, she had been unwilling to even discuss the FARC with Andrés, but the clip opened her mind. The helplessness he had felt for years melted away.

These interviews replaced negativity bias with a more hopeful, human narrative—the opposite of what media often does. It unwound misperceptions about rivals amid conflict and cruelty. Peace in Colombia has not been steady. Scattered armed conflict continues throughout the country, displacing three times as many people in 2021 as it did in 2020. There is still an urgent need for peace, and the Casas brothers continue to create media experiences in support of reconciliation. One piece of information gives Andrés hope: Most people he surveyed years ago wanted peace. Even more do now.

As I learned about Emile's work in Colombia, I was moved but also noticed myself becoming uncomfortable. People in struggle often don't need or want saviors—particularly white, American saviors—to fix their problems. I raised this with Casas, who understood but emphatically defended Emile. Although Casas had joined the lab as a student, he was always treated as an equal. Emile had insisted research be grounded in Casas's experience as a Colombian. "He didn't want an assistant," Casas tells me, "he wanted a partner."

In the camps, too, Emile's credentials didn't matter. "He was just a guy," Andrés says, "not a gringo or professor." In a film later released by Casas and his brother, Emile voices this humility: "I don't think it's my place, as an outsider, to give advice, to tell people what to think. All I want to do is share what I know about the human mind and its ability to change."

But what Casas comes back to most is the uncanny connections Emile forged with ex-FARC members they interviewed. A young girl who had been reluctant to share her story finally spoke with Emile, and then shocked the group with a pitch-perfect rendition of Leonard Cohen's "Hallelujah." An imposing former soldier connected with Emile over a shared love of wrestling, and from there went on to tell his story, excavating emotions he might not have shared in years. Casas marveled at these moments, saying, "He made things come out of people."

Later, I watched some of Emile's interview with the soldier. What strikes me isn't the feeling he draws out of this once dangerous man. It's the emotion he gives, and how he let these conversations change him. "You give me hope," Emile says in tears, "not just here, but for humanity."

After he got sick, Emile insisted on returning to Colombia. Surgery had left his skull so fragile he was required to wear a medical helmet while traveling. "His wife probably wanted to kill me," Casas says. It would be Emile's last trip out of the US, and a fitting one, continuing the work of peace. He found inspiration in Colombia's story, and Casas has continued their shared mission. Each year he hosts an international conference, Neuropaz, where scientists discuss ways to foster peace. The meeting is dedicated to Emile.

To unfreeze fear and hatred, Emile and Casas didn't need to trick people. They simply told Colombians the truth about the other side. As Casas describes, "We Colombians are on the brink of peace. This research teaches us that the best way to achieve it is to think better of others." The same strategy has worked in subsequent studies carried out in the US and elsewhere.

When problems hide under the surface, light can be the best disinfectant. But in our politics, plenty of rot is visible already. What's hidden all around us is a peaceable, inquisitive majority drowned out by extreme voices. Some of our cultural fault lines are based on misunderstanding, and hopeful skepticism can be a powerful tool in mending them. Sunlight—in the form of clear, simple data—reveals something that looks a lot like possibility.

Disagreeing Better

As we've seen, cynical people take preemptive strikes and bring out the worst in others. Hopeful individuals take leaps of faith and bring out others' best. Some of these are grand gestures. Kennedy and Khrushchev's de-escalation was like Project RYaN in reverse: loud trust on a global stage.

Other leaps are small but still powerful, like spending time with people we disagree with. This sounds simple, but it's started to feel impossible. As recently as 2016, 51 percent of Americans said it would be "interesting and informative" to talk with a rival, edging out the 46 percent who said it would be "stressful and frustrating." Five years later, this enthusiasm had evaporated: Fifty-nine percent said these conversations would be frustrating, and less than 40 percent said they'd be interesting. Asked to compare talking with rivals to other activities, both Democrats and Republicans said they'd prefer a painful dental procedure.

In a divided world, bringing people together is not like pulling teeth; it's harder. Chatting with rivals feels dangerous and even immoral—like the Romans inviting Visigoths for a beer during the siege. Even if they could get over their distaste, people don't see the point of cross-party conversations. In 2022, my lab asked hundreds of people what would happen if Republicans and Democrats talked politics. Most believed that people would agree with one another even *less* after the conversation. One Democrat from Pennsylvania wrote, "Political dialogue is doomed." A Texas Republican said, "Civility is dead. Respectfully disagreeing is dead."

They were both wrong. In the summer of 2022, my lab invited over a hundred Americans to set aside their trepidation and join twenty-minute Zoom calls with a rival. After logging on, tidying the rooms behind them, and remembering to unmute, the pairs talked about issues like gun control, climate change, and abortion. Our team had ensured that each person truly disagreed with their partner. We also set up contingency plans for what to do if people insulted or threatened each other.

To everyone's surprise, these conversations were wonderful. People clashed, but also listened. When we asked them to rate the experience on a scale from 1 (very negative) to 100 (very positive), the most common response was exactly 100. After talking with a rival, participants' dislike of rivals plummeted by more than twenty points on a hundred-point scale and remained lower three months later. If the rest of the country joined them, the clock on our nation's partisan aggression would roll backward to the Clinton era: not an entirely peaceful time, but not nearly as vicious as our politics are today. People also left these chats less likely to dehumanize the other side, and humbler about their own opinions.

If social shark attacks scare us away from everyday interactions, they terrify us of rivals. And if conversations with strangers are surprisingly positive, meeting across difference is astonishingly useful. Emile believed this in his bones; one of the last experiments he ever published examined conversations between rivals. He acted on this faith personally as well. Nour Kteily, one of his collaborators, watched Emile interact with conservative friends on Facebook. Night after night, he debated gun control, immigration, and whatever else came up, never relenting his position, but not dismissing rivals, either. "Even when they resisted his message," Kteily remembers, "his empathy blew out the candles of their indignation."

Of course, not every conversation softens conflict. Thousands of Thanksgiving dinners have disintegrated before the pumpkin pie arrives, because people can be just as mean in person as they are online. It's not enough to talk *at* rivals; to be productive, encounters require mental and emotional work. Studies reveal a recipe for disagreeing better:

1. Good disagreers ask questions instead of making statements.
2. They work to get underneath people's opinions to their stories.
3. When they spot common ground, good disagreers name it.
4. When they are unsure about something, they say so rather than pretending to be confident.

These ingredients each decrease the chance dissent will devolve into toxic conflict. But good disagreement is more than nice; it's powerful. In experiments, people given the recipe you see here listened more intently and asked better questions. But the people they talked with also became more open-minded, even though they received no training. Outrage is contagious, but so are curiosity and humility.

Beyond Peace

Conflict engulfs nations, stalls government, and threatens democracy itself. Most people want greater peace. Better data, by revealing all we have in common, can open a passageway toward this goal.

But is that all we should strive for? Some forms of "peace" look a lot like the status quo. Researchers find that when people in historically less powerful groups—for instance, Black Americans—desire harmony with more powerful groups, they are less likely to challenge injustice, bias, and abuse. Many divisions around the world are imbalanced; people in power take land, freedom, and life from those who don't have it. Such oppression is not "conflict," just as a mugging is not a street fight.

I struggle with this. My research reveals that empathy can be a powerful tool for bridging difference and reducing animosity. But lots of people have every reason to feel angry—when their group has been victimized for generations. Sometimes when I speak about compassion, I imagine someone from a marginalized group watching, feeling scolded for not smiling more at people who would extinguish their rights. *What nerve*, they might think, *asking us to sacrifice justice at the altar of peace.*

Hope is a powerful social force. It can strengthen communities, promote understanding, and rebuild trust. But what good is any of that if feeling better stands in the way of doing better? People disagree and dislike one another, for sure. But is that really the worst of our problems? Voter suppression silences millions of Americans. A rising force in our culture is rolling back rights held by women, immigrants, the poor, and the natural world—and threatening democracy itself.

Is hope the sweetener that helps oppression go down easier? More times than I can count, I've worried that my work might be used as a psychological sedative, calming people just when they need to be agitated.

Reflecting on these doubts now, I sense my own cynicism as well. Perhaps human togetherness is too weak and friendly for our tumultuous times. Or maybe not. The only way to know for sure is to follow the science where it leads us.

Section III

THE FUTURE OF HOPE

Chapter 9

Building the World We Want

"The challenge for us ... is not to grieve over social change, but to guide it."
—Robert Putnam

Millions of Americans lived in destitution while oligarchs flashed unthinkable luxury. Zero-sum self-interest ruled the nation. Politicians pursued increasingly extreme agendas and cooperation across parties broke down, sinking the nation into gridlock. New media outlets gawked at the nation's moral collapse, broadcasting a parade of scandals.

This wasn't the 2020s; it was the 1890s. But just like now, cynicism was alive and well. Almost half of America's wealth was owned by the top 1 percent, making the nation nearly twice as unequal as it was in 2022. "Robber barons" like Andrew Carnegie and Leland Stanford monopolized whole industries. "They raised the nation's productivity on the backs of poor, mistreated, and child labor," one historian writes. "The underclass, packed into fetid slums, hated the modern pharaohs ruling over them." The rich despised the poor right back. The financial speculator Jay Gould bragged that he could "hire one-half of the working class to shoot the other half to death."

The elite hadn't gotten there by themselves. Lawmakers helped robber barons acquire their wealth, while social Darwinists helped justify it. "Some people were better at the contest of life than others," one writer claimed.

"The good ones climbed out of the jungle of savagery and passed their talents to their offspring, who climbed still higher." Classism rose, and post–Civil War racial progress reversed. The Supreme Court's decision in *Plessy v. Ferguson* ushered in Jim Crow policies designed to disenfranchise Black Americans. By 1908, Black voting rates across the South had fallen by over 60 percent from their peaks. During the 1890s, a lynching occurred somewhere in the US every two days.

Meanwhile, the telephone, telegram, and daily newspaper flooded people with new information, but not always the useful sort. William Randolph Hearst pioneered "yellow journalism," lurid coverage targeted at audiences' negativity bias. His *San Francisco Examiner* devoted almost a quarter of its stories to crime, old-fashioned clickbait that exaggerated details to entice readers.

People plugged into vast new networks while older connections withered. Millions of Americans had left stable rural communities for cities full of strangers; family shops lost business to mail-order catalogs. In a 1912 presidential campaign speech, Woodrow Wilson mourned our losses: "All over the Union, people are coming to feel they have no control over the course of their affairs... The everyday relationships of men are largely with great impersonal concerns, with organizations, not with other individual men."

Americans living during this time could have easily concluded that the nation was on a one-way street leading to social decay. Yet it wasn't. After a chaotic and painful labor, the twentieth century gave birth to a vibrant progressive movement. Activists, workers, and civic leaders assembled into a vast array of new organizations. They built power through strikes, lobbying, and public engagement, and racked up staggering accomplishments. Here are just some of the policies passed between 1888 and 1920: women's suffrage, the income tax, the creation of the Food and Drug Administration, Federal Trade Commission, and National Forest and Parks System, child labor laws, the eight-hour workday, campaign finance regulation, and public kindergarten.

Alongside material change, people started to think differently. An ideology of Christian Social Gospel spread, emphasizing moral responsibility to help those in need. This philosophy overtook social Darwinism. An analysis of all published books from the time finds that as the term "Social Gospel" ascended in use, the phrase "survival of the fittest" dwindled.

So did cynicism. The General Social Survey began in 1972, when nearly half of Americans believed that "most people can be trusted." That's where we started this book, but it's not the first time scientists had asked this question. In 1960, 58 percent of Americans had trusted their fellow citizens. During World War II, the number was an astonishing 73 percent. As the century reached its middle, the nation's mind was turned toward goodness, its heart toward community, and its eye toward progress.

This high-water mark for trust sounds like a fantasy today. It would have also sounded like wishful thinking decades *before* it occurred. After learning about the trust recession, I began to wonder if the opposite had ever occurred. Where and when did people's faith in one another build instead of break? Those stories, I thought, could provide clues as to how we might escape the cynicism trap now.

It turns out that positive change happened right here. Before American community disintegrated, it strengthened. Before things got worse, they got much better, in what political scientist Robert Putnam calls "the upswing."

It's easy to forget these accomplishments because progress has a habit of hiding its tracks. New generations take for granted rights and abilities their ancestors dreamed of. Cynicism takes root in our amnesia. Psychologically, the country is worse off than decades ago. This period of decline dominates our minds, and for good reason: Unless you're past retirement age, your entire life has occurred within it. But zoom out a bit, and the past tells a different story. Fierce, hopeful work has changed the world before and can do it again. The rest of this book will imagine what that could look like.

If we wanted to follow the path of earlier Progressives and reduce cynicism on a grand scale, a powerful starting point would be to decrease inequality. But ironically, one barrier that stands in the way of this goal is cynicism itself, and how easily we point it at the poorest among us.

Given Less and Trusted Less

In 1976, Ronald Reagan began telling a story, one he would repeat during his failed presidential run, then his successful one, then throughout his first term. "In Chicago, they found a woman who... used 80 names, 30 addresses, 15 telephone numbers to collect food stamps, veterans benefits for four nonexistent deceased veteran husbands, as well as welfare." Depending on the speech, he might mention her Cadillac and furs, or the fake kids she used to swindle the system. Journalists mocked Reagan for exaggerating, but the real story was even worse. Linda Taylor, the woman in his speeches, really did use dozens of aliases to steal hundreds of thousands of dollars in public benefits. She also trafficked children and was credibly accused of murder.

Taylor abused many people and was a victim of abuse as well. Her life was a complex tragedy. But in Reagan's hands, it became a cartoon: the "welfare queen." According to him, catching one Linda Taylor meant there were countless others around the country, buying caviar with food stamps and pumping out children to live off taxpayers' sweat. Reagan used nut picking—citing extreme examples and pretending they're the norm—to convince people that the most corrupt welfare recipient represented them all.

By 1978, just two years after Reagan introduced the welfare queen story, 84 percent of Illinois voters thought welfare fraud was the most pressing issue facing their state. Nationally, fraud investigations jumped by more than 700 percent between 1970 and 1980. America spent millions of dollars to hunt down cheaters and found barely any. That's because most recipients of public aid are nothing like Linda Taylor. A 2018 report found that for every ten thousand households that participated in the Supplemental Nutrition Assistance Program (SNAP), only fourteen included fraudsters.

Welfare queen myths were inaccurate, offensive, and effective: They inspired public support for a 1982 bill that cut $25 billion from social welfare programs, a massive transfer of wealth upward. That year alone, more than a million Americans lost access to food stamps, many cast into hunger. Public benefits in the US continued to decrease steadily over the decades. Between 1993 and 2018, the real value of cash provided by the Temporary Assistance for Needy Families program was cut by nearly 80 percent. Over the same period, the number of American households living in extreme poverty—defined as living on less than $2 per day per person—soared, more than doubling by some estimates.

Racist, sexist tropes like "welfare queens" live on. White Americans estimate that 37 percent of US welfare recipients are Black, nearly double the actual figure of 21 percent. And the more public aid white people believe goes to Black recipients, the less they support welfare in general. During the COVID pandemic, the government broadened some welfare programs. This led to an avalanche of criticisms, anchored in the fear that poor people would take advantage. Republican congressman Matt Gaetz called public aid recipients "couch potatoes"; Democratic senator Joe Manchin worried parents would use child tax credits to buy drugs.

Stereotypes like these build on a widespread and long-standing suspicion of people who lack resources. Given less, poor people are also trusted less, a psychological inequality that reinforces the economic kind.

A Constitution for Knaves

Two hundred years before Reagan's time, the Scottish philosopher David Hume laid out his recipe for a society: "Every man ought to be supposed a knave, and to have no other end, in all his actions, than private interest. By this interest we must govern him, and, by means of it, make him…co-operate to public good."

These suggestions were simple and perfectly cynical. People are selfish, and society can function only if they are browbeaten into pretending not to be. Jack Welch built corporate worlds to suit *homo economicus*; Hume suggested we build government for him as well.

A nation built on Hume's ideas would start with a "constitution for knaves." Rather than guaranteeing liberty, this document would reduce it. East Germany under the Stasi spied on and threatened citizens to keep them in line. But even in freer nations, some people live under a constitution for knaves—usually those with the fewest rights, opportunities, and resources. In fact, if you want to guess how marginalized someone is, a useful sign is how cynically everyone else treats them.

One person who lived that story is William Goodwin. He was born in West Oakland ("before it was gentrified," he tells me). His father died when he was young, and his mother struggled to support him and his siblings. From the age of six, he washed neighbors' cars and mowed their lawns. When his mother was low on funds, the corner store ran a credit line so the Goodwins could eat. The family was enterprising, faith-driven, and supported by community. "We didn't know we were poor," he remembers.

When William reached fifth grade, his mother petitioned to get him a better education. He took tests in grammar and geometry and was soon sent to a new school perched in the Oakland Hills. By the time he was in high school, he woke up at 5:00 a.m., took two buses, and walked through a parking lot full of other students' gleaming new cars. If he hadn't known he was poor, he did now. Goodwin was out of place at his new schools, and out of place when he got home, too. "He thinks he's somebody now," friends would say, mocking him as too good for his old neighborhood.

Soon, Goodwin began running the school's store and discovered a passion for business. After graduation, he worked in sales at Levi's, then became an underwriting liaison for a major insurance company, climbing the ranks for over a decade. As a student, he had felt like a foreigner in the Oakland Hills. Now he lived there, pleased to support his mother through her last years of life, and later his young daughter through her first.

It all came crashing down when Goodwin developed a degenerative nerve condition. A needling sensation took root in his neck and barely ever left. His back spasmed constantly. He had trouble sitting and keeping his neck and head upright. Goodwin wanted to keep working but needed accommodations. In his department, each liaison was responsible for processing eighty insurance claims a day, rapid-fire work that also meant being anchored to a desktop nearly every minute. Numbers meant everything, and Goodwin simply couldn't make his anymore.

A standing desk, longer breaks, and reduced quota all could have helped Goodwin stay at the job he loved. Instead, his company busted him down to the filing room. In a nod to his condition, the demotion did not come with a pay cut. As a result, Goodwin's colleagues murmured that he was faking his symptoms to do easier work.

Black and now disabled, Goodwin arrived at an intersection of American stereotypes. People he had known for years ignored his long-standing work ethic and cast suspicion on him. Goodwin's boss made him visit an army of doctors to verify his condition was real, then tried to force him into a risky surgery that helps only 50 percent of patients. When he refused, Goodwin was laid off and began applying for disability.

The process lasted the better part of a year and seemed built to demoralize. He would spend half a day waiting for a ten-minute conversation with a benefits counselor, then be told he had to visit a different office, only to be sent back to the first. He pinballed from one doctor's appointment to another, each time answering the same questions. Was he *really* sick? Did he *really* want to work at all? "The pressure beamed down on me," Goodwin remembers, "reinforcing the same old stereotypes, a Black man trying to get one over."

Adding insult to literal injury, he felt "helpless, alone, and distrusted, even though I was telling the truth." Eventually, the system's doubts filtered into his mind. "Am I who they think I am?" he wondered one day, staring out the window of an Oakland train. "Will I end up *being* who they think I am?"

His benefits application was rejected. He appealed, and then "things got really technical." Interrogation-like interviews, appointments, and forms multiplied. He was rejected again and had only so many appeals left. He thought back to the poverty of his childhood. *I'm going back there*, he thought. *I can't believe it.* Goodwin considered giving in, but someone was counting on him. "I thought about my daughter, and how I could be the best father to her. Injury or no injury, I had to put all my focus in her."

He appealed again, this time hiring a lawyer, who presented the same evidence of disability Goodwin had included in his first two applications. This third was approved—for a term of three months. Goodwin estimates he spent seventy-five hours on the process. In the meantime, court officials and employment doctors got paid, as did his lawyer—who took 15 percent off the top of Goodwin's award. Then the process started all over again as he applied for long-term disability.

Meanwhile, Goodwin's financial life unraveled. His car broke down and he didn't have the money to fix it, leaving him unable to seek out work he could do with his condition. He and his daughter moved in with friends; they were, as he puts it, "homeless even though we had somewhere to live." It would be years before he finally received long-term benefits, and some stability.

For his daughter's sake, Goodwin needed to believe in himself even when no one else did. What he stopped believing in, though, was the system. He watched his boss look for any excuse to drop a disabled employee, "like a landlord trying to evict a tenant." Government programs, he says, "aim to keep you where you are, not pull you out, so the narrative stays alive and they can stay employed." Having been judged, Goodwin judged right back.

Trust for the Few; Cynicism for the Many

According to the US Census, nine million American children experienced poverty in 2022. Childhood poverty is a moral catastrophe, and an economic one. A report from the Center for American Progress estimated that it costs the US $500 billion per year because poor children grow into adults who work less and use more health care than others. Poverty hurts Americans at the beginning of life and hurries its end. A 2019 report found that the poorest US citizens die an average of five years earlier than the UK's poor, whereas wealthy Americans live just weeks less than their British counterparts.

As we've seen, periods of high inequality raise cynicism. But the reverse is also true: Cynicism feeds inequality. At least some studies suggest that a person's faith in fellow citizens predicts whether they will support public aid for the poor. When a nation stops trusting itself, its suspicions first turn toward those with the least.

"Povertyism," discrimination against the poor, is alive and well in stump speeches and government processes. And when the poor seek aid in the US, they are sentenced to death by a thousand paper cuts. A single mother in Louisiana must complete a twenty-six-page application to qualify for food stamps. Journalist Annie Lowrey describes some of its highlights:

> Page three lets her know that she needs to collect paperwork or data in up to 13 different categories—pharmacy printouts from the past three months, four pay stubs, baptismal certificates, proof of who lives in the home...page seven outlines the penalties if she misuses her benefits by, for example, spending them on a cruise ship or at a psychic. Page 15 asks her to detail her income from 24 different sources.

To access food stamps and health care, poor Americans must wait hours to answer demeaning questions that have nothing to do with their requests ("on what date was your child conceived?"). These rituals tax the time of poor Americans, and their minds. Some research finds that scarcity—the feeling of struggling to make ends meet—decreases human mental capacity as much as pulling an all-nighter. Except unlike all-nighters, scarcity follows us for weeks, months, or years.

William Goodwin needed the skills and knowledge he gained in the insurance industry to fight for his benefits. Not everyone has those advantages, and many don't receive the benefits they are owed. In *Poverty, By America*, sociologist Matthew Desmond reports that each year poor Americans pass up over $13 billion in food stamps, $17 billion in tax credits, $38 billion in supplemental security income (SSI), and $62 billion in government health insurance. Many manage to receive benefits only when they hire lawyers, like Goodwin did, collectively paying some of their aid back into the system.

This is not the sign of a policy failing; it's the sign of one working—to keep the poor where they are. Rather than admit to this, legislators claim their prickly rules are essential to stopping cheaters. Benefits applications are written for the world's Linda Taylors—as rare as they may be. Healthcare reimbursement forms are written for people who are faking. Our government invests countless hours and dollars to sniff out a tiny percentage of cheaters. This makes awful economic sense but provides a cynical cover for mistreating the poor.

Our system lobs preemptive strikes at anyone who needs help, and the poor reciprocate that contempt. Trust has plummeted across the US, but when inequality rises, faith in people and institutions falls most quickly among lower-income citizens.

The poor live under a constitution for knaves. Well-off Americans live in a far different nation; its founding documents assume the best of them, and its coffers open for them through an array of public benefits. My wife and I can contribute to a "529" college account for our daughters. Earnings on this investment are not taxed, meaning we keep more of our income than we would without the 529 program. In other words, we receive thousands of dollars in public aid each year. To benefit from this, I don't have to march my children to a government office, wait hours to be seen, or endure questions about why my hair and skin are darker than theirs. People who hold mortgages can write off the interest they pay against their taxes, again subsidized by the government. In 2020, those tax breaks dwarfed all spending on housing assistance for the poor by a ratio of nearly four to one. If you own a home, you receive much more public aid than if you need one.

In a 2010 study, 64 percent of parents who put money in 529 accounts wrongly reported that they had "not used a government social program." The government gives this money away so easily that two out of every three people who receive it don't even notice. Wealthy Americans live in a welfare state invisible to nearly everyone, including themselves.

The American poor learn quickly and often that our nation sees them as shiftless cheats. Those with means learn the opposite, plied with benefits and the benefit of the doubt. In the most rarefied air, billionaires pay a lower percentage of their income in taxes than nearly any other income bracket. They also earn outsized influence, empowered through the passage of Citizens United to funnel massive amounts of money into political campaigns.

In the world's richest nation, millions of people drink water that fails to meet federal health standards. The *American Journal of Public Health* estimates that between thirty-five thousand and forty-five thousand Americans die from curable illnesses each year because they lack health insurance. Medical and other debt also significantly increase the chance people will attempt suicide. Meanwhile, the US dominates the world market for "superyachts," accounting for about a quarter of global sales.

It's hard to stomach being the villains of a slow-motion moral catastrophe. There are at least two ways we can respond to this dissonance. The first is to shut ourselves off, trying not to think about people who have less than we do. If we do think about them, we can turn up our cynicism to justify why they must suffer. Social Darwinists claim the rich are better competitors than everyone else. An even older prejudice claims the rich are just plain better. Aristotle believed aristocrats were society's frontal lobes—guided by principle, conscience, and forbearance. The wretched masses were greedy *homo economicus* who must be saved from themselves. Slaves were better off in chains, for they would squander freedom.

We revive this prejudice every time we suspect the poor will run off with public aid. This all contributes to a system of cynicism for the many and faith in the few: a philosophy Diogenes would despise, and the data contradict, but that provides psychological armor for an unjust world.

Redistributing Trust

There's a second way we could respond to the crisis of inequality—one that is more compassionate and data driven. This alternate path promises civic strength and moral health, but also requires painful inner work for those of us with privilege. To pursue it, we would need to grapple with the ways we benefit from the immiseration of our neighbors, and how we could contribute to a more just system.

That change is probably more popular than you think. Common Ground, the survey that identified issues on which Americans agree, found that more than 80 percent—including two-thirds of Republicans—support the expansion of food stamp programs in at least some cases. Over 70 percent—including more than half of Republicans—support expanding prekindergarten for poor children.

Most people want social systems to support those in need. What would that look like? One starting point would be to write policy as if most citizens are who we hope they are, not the cheats we've been taught to fear. It would mean ripping up the constitution for knaves and giving everyone, rich, poor, and middle class, a chance to prove themselves.

In 2016, William Goodwin received such a chance. A friend told him about UpTogether, a national organization that provides aid to poor

Americans. Where public programs confuse and humiliate, UpTogether is built to empower. Its members set their own financial goals, and the organization gives them money and other resources to pursue them. This approach is often called "direct cash transfers." No twenty-six-page forms, no micromanaging or red tape. Suspicion replaced with simple, loud trust.

At first, Goodwin didn't trust back. "You hear about a program giving out resources; it sounds like a scam. I kept the money in the bank for a minute, waiting for the other shoe to drop." Lots of members start out that way: doubting anyone who doesn't doubt them. UpTogether's CEO, Jesús Gerena, sees pain under their distrust. "After all that neglect, people think, *No way am I worthy of recognition, let alone investment.*" Gerena remembers giving a laptop to one UpTogether member, who asked where to sign it out—sure he'd need to bring it back, in disbelief that it was really his. "Just don't call us if it breaks," Gerena joked.

With UpTogether's help, Goodwin set goals to pay off his car loan and start planning his daughter's education, sending his daughter on a university tour. After that trip, "She was different," he remembers. "Higher education became more than an idea. It was real to her, and that made it real to me." Goodwin now serves on the board of multiple nonprofit organizations, and he works in partnership with his county to advocate for housing equity. His daughter attends college just east of the Bay Area. He loves having her close by, even if he wishes she'd visit more often.

The Goodwins changed their story, and then William became more ambitious—about helping others. "At first it was like, 'We've got this money, how do we spend it?' but then I thought, 'How do we give back?'" He joined a group of UpTogether members to revamp the computer room of a local community center, and founded a literacy program at the East Oakland Library that is still running five years later. "I can't seem to stop," Goodwin tells me. "There's a responsibility on me, as someone receiving [cash transfers], to make good on them."

As we've seen, leaps of faith inspire people to step up and earn our trust.

Given a chance, UpTogether members rise. Within two years of joining, the average UpTogether family reports an increase of more than 25 percent in income, and a 36 percent *decrease* in their dependence on government assistance. Over 80 percent of children in UpTogether families receive good or improving grades in that same period.

UpTogether might seem like a local feel-good story—nice for Goodwin; impossible to spread. But cash transfers are an old and popular idea. Plenty of liberals support them, but so did archconservative economist Milton Friedman, who called them a "negative income tax." As he argued in 1962,

> A negative income tax is a proposal to help poor people by giving them money, which is what they need, rather than as now, by requiring them to come before a government official to tally all their assets and liabilities and be told that you may spend X dollars on rent, Y dollars on food, etc.

To liberals, cash transfers support justice. To Friedman, they strengthened the free market. When people who are so far apart agree, the rest of us might want to pay attention. And the more scientists examine cash transfers, the wiser they appear. An explosion of new programs has delivered transfers to poor people—first in the Global South, but increasingly in Europe and North America as well. Researchers have followed, marking the effects of these programs over time.

The results undermine our popular, cynical stories. When the poor receive cash, they don't fritter it away. In nearly twenty studies across Asia, Africa, and Latin America, scientists found no evidence that people use transfers on "temptation goods," such as alcohol and tobacco. In 2018, the charity Foundations for Social Change gave a no-strings-attached transfer of $7,500 to each of fifty unhoused individuals living in Vancouver. People who received this windfall spent no more on temptation goods than those who had not received it. Instead, they used their new funds to pay for food, clothes, and rent. Recipients were more likely to find stable housing and *less* likely to use public services.

A 2023 study of the program calculated that each cash transfer saved the shelter system $8,277. In other words, trusting Vancouver's poorest residents was not just moral, it was economical. Across the world, people tend to spend cash transfers wisely. In agricultural societies, families invest in livestock and tools, and earn more from farming two years later.

But don't transfers encourage people to stop working? Like other stereotypes about the poor, this one breaks to pieces against the data. Studies in South Africa, Canada, and the US found no major reductions in labor among people who received transfers. There is one exception: When families receive money, parents can spend more time with their kids, creating virtuous cycles that spread across the generations.

When adults receive cash transfers, their children are more likely to enroll in school and aspire to college like William Goodwin's daughter. As adults, they earn more than children whose parents received no transfers. These benefits go under the skin as well. Low-income children are more prone to mental illness but are protected if they live in states that offer larger cash benefits. And when a new program gave money to poor mothers in Pennsylvania, their infants' brains developed more rapidly, possibly because moms were able to create a richer environment for them.

If you're surprised by these studies, that's unsurprising. The researchers who gave cash transfers to unhoused Vancouverites also asked more than a thousand Canadians what would happen to the money. Most predicted it would go toward drugs. I don't think the people who made these guesses hate unhoused individuals. If in the past you've thought poor people would waste cash transfers, I don't imagine you are full of hate, either, in part because I shared these stereotypes. Before reading the research on cash transfers, I would have worried that their recipients might waste the funds or work less after receiving them.

I'm not proud of my past prejudice but realize it's another common side effect of our culture's cynicism. We should arm ourselves with more skepticism and turn it inward as well. Where do our assumptions come from? What purpose do they serve? Who else do they benefit? Cynical beliefs uphold the status quo. If the poor are knaves, the wealthy have no reason to question their advantages. Programs like UpTogether and Foundations for Social Change upend these views. Their hopeful experiments show us what can happen when society instead takes a leap of faith on poor people.

Earlier this decade, we did. During the pandemic, the US government sent direct cash transfers, rent and childcare assistance, and other public support at a level unseen in decades. In the face of a generational disaster, poverty in America *decreased*. This was a massive victory for the poor, and for national decency toward all citizens. It could have been celebrated by solutions journalism, heralded as a sign of what could be. Instead, the moment ended, public support receded, and extreme need skyrocketed again.

To do better, we can change our narratives about the poor, using media and education for asset-framing instead of cynicism. As those stories rise, so might compassionate, creative policies to lift people up.

Redistributing trust also means considering who receives it too easily. Matthew Desmond argues that nearly all extreme poverty in the US could be "canceled" with about $177 billion per year. The top 1 percent of wealthy Americans evade $175 billion per year in taxes. In other words, an enormous amount of suffering could be offset, not by increasing top tax rates to what they were in higher-trust times like the 1950s (91 percent) or the 1970s (70 percent)—but by just getting our richest citizens to pay what they owe.

If we ask fewer questions of poor people, we might ask more questions of the rich, for instance, keeping a closer eye on their tax payments to constrain cheating. Preventing dark money from flooding into politics and forcing disclosures of super PAC donors could invite public scrutiny of their tactics.

Of course, cash alone will not bring about the social progress many of us want. A tapestry of injustice and oppression covers vulnerable people. Change-makers around the world now are fighting these forces, wielding hope as one of their most effective tools. Like the twentieth-century Progressives before them, they are breaking down doors. Like people back then, we might be surprised by all they can accomplish.

Chapter 10

The Optimism of Activism

In 1967, Martin Luther King Jr. gave a speech to the American Psychological Association and politely dismantled its priorities. When someone thrives, psychology calls them "well-adjusted." When a person acts out or fails to launch, they get pegged as "maladjusted." This antiseptic insult makes it sound like anyone who has a problem *is* the problem. King discarded this idea:

> There are some things in our society, some things in our world, to which we should never be adjusted... We must never adjust ourselves to racial discrimination and racial segregation. We must never adjust ourselves to religious bigotry. We must never adjust ourselves to economic conditions that take necessities from the many to give luxuries to the few. We must never adjust ourselves to the madness of militarism, and the self-defeating effects of physical violence.

Instead of trying to be well-adjusted, King said, we should pursue an alternative. "Our world is in dire need of a new organization," he proclaimed, "the International Association for the Advancement of *Creative Maladjustment*." Creative maladjustment, he explained, is a moral restlessness triggered by wrongdoing that propels social change.

Throughout this book, we've explored the power of hope to improve lives, relationships, and communities. But what if hope is just another path toward being well-adjusted, ignoring our many problems? Injustice, inequity, violence, and cruelty are real, and they can't be wished away by the power of positive thinking. Perhaps cynicism is a badge of moral clarity.

This makes sense on the surface, but just like the cynical genius illusion—the idea that people who trust less are more clever—it falls apart when we look more closely. It actually *is* hope—the sense that things could improve in the future—mixed with fury, that inspires people to fight for progress, even when victory seems well out of reach.

The Impossible Will Have to Wait Awhile

Václav Havel began the 1980s in a tiny cell at Prague's Ruzyně Prison. He had grown up well-heeled and well-fed, shuttling between family homes in Prague and the countryside. Then Communism rose, leaving members of the middle classes with fewer options for work and education. Havel channeled his mind into the arts, crafting plays that mocked and criticized the Communist regime. They were a hit, and the young writer became friends with Samuel Beckett, Kurt Vonnegut, and a generation of Czechoslovak artists. In 1968, Havel and other activists joined the Prague Spring, a peaceful movement to relax Communist rule around the country.

Czechoslovakia appeared to be marching forward into a brighter future, and then was abruptly yanked backward. Hundreds of thousands of Soviet-allied soldiers flooded into the nation and put an abrupt end to the Spring, along with many of the nation's dreams. Surveillance and violence escalated. Any public dissent against the government could cost someone their job. Travel to and from non-Communist countries was restricted.

Havel had not planned to enter politics but found it impossible to stay silent. In 1978, he wrote *The Power of the Powerless*, which describes how oppressive governments steal people's hope. In the essay, he imagines a grocer who hangs a Communist slogan in his shop to avoid persecution.

Neighbors know he doesn't believe the slogan, so the sign actually advertises the grocer's surrender. Soon, neighbors hang their own signs. Everyone knows everyone else is lying, and no one can count on anybody. "By exhibiting their slogans, each compels the other to accept the rules of the game," Havel writes. "They are both victims of the system and its instruments."

Havel refused to remain complicit. He joined Charter 77, a group of dissidents that advocated for a freer Czechoslovakia. He wrote and spoke out against the regime. For this activism, Havel's plays were banned across the country. He was harassed regularly by secret police, and imprisoned several times, his longest sentence spanning from 1979 to 1983.

An oppressive regime dominated Prague. Activists fought it and were crushed. Authoritarianism had won; people had lost. Surveying this from his cell, Havel could have easily concluded that matters would only worsen, with Czechoslovak freedom fading from memory like a dream in the morning. What would that kind of thinking have done to him?

Cynics call out injustice wherever they see it, but that doesn't mean they are change-makers. In surveys of tens of thousands of people across dozens of countries, people who trust others are more likely than cynics to vote, sign petitions, join lawful demonstrations, and occupy buildings in protest. Cynicism tunes people in to their culture's illnesses but makes any cure seem impossible. It whispers (or screams) that our government is abusive because *every* government is, that this politician is corrupt because they *all* are. If that's true, trying to make a difference is delusional. Like Havel's shopkeeper, cynics give up and give in, sending a clear signal to everyone else: If you want to cause trouble, don't expect me to have your back.

In other words, cynicism is a tool of the status quo. Autocrats encourage it for exactly that reason. In 2016, the RAND Corporation analyzed Vladimir Putin's propaganda operation. Russian disinformation is a steady "firehose of falsehoods" sprayed through state television, social media, and newspapers. But RAND discovered an unexpected twist. Most repressive states keep a tight control on information. Their propaganda might be untrue, but it's consistent. Russia didn't follow that recipe; Putin changes his message at will. At one point, he insisted Russia had no interest in Crimea and no troops stationed there. But before long, he admitted ordering troops into the region and said that it should join Russia.

Why wouldn't Putin control his messaging? Perhaps his mission isn't to convince people of anything. In 2021, researchers interviewed Russians to examine the effects of "unconvincing propaganda," which doesn't bother to be credible or consistent. After watching state media, citizens were disgusted, but also hopeless about politics in general. "I don't need to know about [political issues], it's useless," said one. "I don't see any reason to get involved or care about politics," said another.

The firehose of falsehoods was designed to wear down citizens' sense of reality. As the philosopher Hannah Arendt wrote, "The aim of totalitarian education has never been to instill convictions but to destroy the capacity to form any." Cynicism leaves people in a dark sort of complacency.

Creative maladjustment is different. It shares, with cynicism, a diagnosis that something is wrong. But whereas cynics find that cold and tiring, changemakers find it fiery and energizing: not because things will get better, but because they *could*. Havel lived this principle even from his cell. Writing to his wife, Olga, he reflected, "Hope is a dimension of the spirit. It is not outside us, but within us. When you lose it, you must seek it again WITHIN YOURSELF and in people around you—not in objects or even in events."

During his time in prison, Havel continued corresponding with members of Charter 77. Meanwhile, the Czechoslovak economy worsened, and its citizens grew emboldened to demand change. As the movement snowballed, people realized that the Communist regime was vulnerable, which gave them even more energy to challenge it. Havel had predicted this pattern years earlier in *The Power of the Powerless*:

> The crust presented by the life of lies is made of strange stuff. As long as it seals off hermetically the entire society, it appears to be made of

stone. But the moment someone breaks through in one place...the whole crust seems then to be made of a tissue on the point of tearing.

In 1989, Czech totalitarianism was shredded by the "Velvet Revolution." That November, police violently suppressed a protest in Prague, triggering an avalanche of creative maladjustment. Students went on strike and theaters read proclamations against the government instead of performing their plays. With radio and television under government control, people hung homemade posters demanding change. It was a photonegative of the grocery store propaganda Havel had imagined. In risking their safety to protest, Czech people realized how many others were on their side.

Havel and Charter 77 capitalized on the moment, creating the Civic Forum, a pop-up organization that served as the unified voice of the Czech people and spearheaded the growing movement. Civic Forum leaders called for the firing of police officers who had attacked protesters and organized a general strike, which was supported by three-quarters of the population. Protests swelled to tens of thousands, then hundreds of thousands of people. Within two weeks, Communist rule in Czechoslovakia had ended without any large-scale violence. Havel began the 1980s in a prison cell. Three days before the decade ended, he became his country's first democratically elected president.

Countless people might feel the way Czechs felt then. Democracy is on the ropes; elite abuse is ascendant. We might decide that positive change is impossible now, too. Skepticism tells us something truer: The future materializes second by second, and we have a hand in shaping it. But what inspires us to take control?

The Emotional Alloy of Change

Emile studied peace, but he also fought for change throughout his life. After President Trump issued a "Muslim ban" in 2017, Emile headed to Philadelphia International Airport to protest in solidarity. He and his kids, Clara and Atticus, joined demonstrations for the rights of Black Americans, the LGBTQ community, and the environment. Teaching at an elite high school, he challenged students on issues of race and class.

In 2000, he gave a sermon about his work at a Unitarian Universalist church near his hometown. Activism, he explained, was noisy, plodding, and inconsistent. Human rights move forward, then backward. Sometimes his high schoolers listened; sometimes they didn't. "Every year, very similar problems, similar issues, similar arguments," Emile said. The work reminded him of the mythical figure Sisyphus, condemned by the gods to push a boulder up a hill, only to watch it roll back down.

Wouldn't this toil and repetition extinguish any activist's fire? Emile didn't think so. "I've had a growing realization about Sisyphus...," he told the congregation. "His story only becomes a tragedy if he doesn't like pushing the boulder—or if he is fixated on the summit." What if he, like Havel and activists around the world, exerted themselves not because they were sure to reach the summit, but because it was the right thing to do? There could be meaning in the struggle. "The premise is set: The boulder must be pushed," Emile announced. "Whether it is a comedy, a drama, or a tragedy is up to the one pushing."

When it comes to social movements, who keeps pushing, and who abandons their boulder? In 2022, researchers reviewed the science of "collective action" such as protests and boycotts. Examining data from over 120,000 people in dozens of nations, they discovered two emotional forces that drive collective action. People take part in social movements when they feel *righteous anger* at injustice, and when they experience *efficacy*, a sense that they can do something about it.

Efficacy without anger can leave us complacent. Anger without efficacy leaves us paralyzed and cynical. Neither inspires much action. But together, they form an emotional alloy of social change, which looks a lot like creative maladjustment.

Plenty of us are already outraged about plenty of things. Efficacy can be harder to come by. How can we cultivate it in difficult times? One ingredient is the belief that others will step up. In the 1960s, researchers asked Black Americans about their willingness to take part in sit-ins against segregation. A Black person who thought whites supported racial progress was about 20 percent more likely to protest than one who believed whites were unsupportive. Whites who observed sit-ins, in turn, *became* more supportive of racial justice afterward, and more likely to join in later protests. This created a virtuous self-fulfilling prophecy: Black protesters who believed others would join in took action, and influenced them to do exactly that.

Authoritarian elites count on citizens counting each other out, becoming victims of both the system and its instruments. But this also means that when brave people step into the light—often at great risk to themselves—they can produce enormous change. In 1988, only 12 percent of Americans supported same-sex marriage. In the decades that followed, a rising number of gay and lesbian people came out. These courageous choices exposed them to bigotry but also raised visibility. Galvanized in part by the AIDS crisis, LGBTQ activists grew more forceful in demanding rights. By 2015, the nation had flipped—60 percent of Americans supported same-sex marriage and the Supreme Court ruled it should be legal across the country, making this one of the fastest-moving political issues in the nation's history.

Changes such as this look like miracles but are really more like math. Scientists have found that when at least 25 percent of people consistently champion an idea or moral movement, it is much more likely to catch fire. More research on this topic is needed, but the science connects with a common story. Activists toil for decades. They groan but keep pushing their various boulders. And then, all at once, the impossible happens. A torrent of support overwhelms the status quo.

This doesn't mean the work of change should fall at the feet of oppressed people, while the majority wait on the sideline for things to be safe. One strategy for bringing bystanders along is to replace cynicism with data. When people believe others are fine with the status quo, they're more likely to remain passive. If they knew what others really felt, they could act together—ripping the signs off their storefronts all at once.

Consider Saudi Arabia, a country in which only about a quarter of women are employed, and fewer work outside domestic settings. A tradition of "male guardianship" laws meant that until recently, men had decision-making power over their wives. Female employment might, then, reflect male preferences. Except it doesn't. In a 2018 study, more than 80 percent of Saudi men believed women *should* be able to work outside the home. They also thought, wrongly, that far fewer men agree with them. Scientists then showed some Saudi men how others really felt. This knowledge gave them peer permission to express their true beliefs. Months later, their wives were nearly twice as likely to have applied for jobs than wives of men who hadn't seen the data, and nearly five times as likely to have gotten interviewed. This is one tiny step toward women's rights in a nation that continues to deny them, but demonstrates that progress can occur through better information.

As we've seen throughout this book, our beliefs skew negative. The truth, then, tends to be a pleasant surprise. Whatever your issue, you might think you're on an island, bristling alone against oppression, while most people don't care. That's probably wrong. Knowing the truth can lend efficacy to your anger and creativity to your maladjustment. It can help us find solidarity with others, pushing together until the boulder—at last—stays atop the hill.

Everyone's Miracles

Stories of creative maladjustment tend to collide with greatness. Václav Havel fought against impossible odds in Prague and won the presidency. Nelson Mandela did the same in apartheid South Africa. Malala Yousafzai risked her life to champion women's education in Pakistan and beyond. Their lives inspire countless others but also feel distant from our own.

Learning about them, ordinary people might conclude that change is driven by superhuman figures who bend history while the rest of us are tossed about in its currents. And if most of us can't make a difference, why bother trying?

One issue that makes me feel this way is voter suppression. In 2013, the Supreme Court voted to roll back sections of the 1965 Voting Rights Act, giving states more leeway to shape elections without federal oversight. In the six years following the decision, state governments closed more than fifteen hundred polling places, while also restricting access to mail-in ballots and purging the rolls so that people must reregister or lose their rights.

These changes make voting harder. Others make votes count for less. Every ten years, states redraw lines between congressional districts. What should be a neutral process has become a political weapon through partisan gerrymandering. Lawmakers from one party design unnatural districts to disenfranchise rival voters. "Packing" refers to quarantining voters into a small number of districts so they influence fewer elections. "Cracking" means spreading rivals into many districts so the opposition forms a minority in each. These tactics allow politicians to choose their voters instead of the other way around, and roll out unpopular policies because they no longer fear being voted out.

I find voter suppression an exhausting issue, in part because it can happen quietly. Most people don't tune in to stories about local secretary of state elections. A majority of both Democrats and Republicans oppose gerrymandering, yet this process is often shrouded from public scrutiny. In these shady corners, our democracy is being dismantled by a well-oiled, well-funded political machine.

Gerrymandering reliably feeds my cynicism. In an attempt at hopeful skepticism, I looked up this issue on Solutions Story Tracker—the website David Bornstein and Tina Rosenberg created to highlight asset-framed news. There, I learned about Katie Fahey. In 2016, Fahey was a twenty-seven-year-old program officer for the Michigan Recycling Coalition. An independent, she cared about national issues but paid closer attention to her civic backyard. Commutes between Grand Rapids and Lansing gave her hours each day to listen to Michigan NPR. As the election loomed, she was "jazzed" about the battle for the position of County Drain Commissioner—the person in charge of water management. If this seems wonky, remember that less than two hours away, the city of Flint had its water poisoned following a series of negligent decisions by local politicians.

Fahey had learned about gerrymandering in fourth grade and been outraged ever since. "All the time," she tells me, "you would hear, 'The people of Michigan want this, but the legislature isn't doing anything about it.'" This racked her with "existential dread," but during that election season, she noticed a change in the people around her. At a nephew's birthday party, friends and family discussed the differences between Bernie Sanders's, Hillary Clinton's, and Donald Trump's childcare policies. Her passion for political detail seemed to be spreading.

After the election, it all turned sour. Division skyrocketed and everyone seemed to think their rivals were evil. Fahey wondered if there was a way to channel people toward an issue they agreed on. She logged on to Facebook one evening and the app pinged her with a memory: Years ago that day, she had posted a complaint about partisan gerrymandering. No one had responded. On a whim, she posted again: "I'd like to take on gerrymandering in Michigan, if you're interested in doing this as well, please let me know ☺."

Unlike the last time, the post attracted a small but committed group, enough to make Fahey think, *I'm not alone here.* Dozens of people replied or sent messages thanking her for doing something about gerrymandering, even though she hadn't done anything yet. *Oh crap,* she thought, *there are thousands of us who want this, all waiting for someone else to do something about it. Maybe we can be those people.*

She started where any of us might, by googling "how do you end gerrymandering?" The search pointed out three options: a lawsuit (which would be temporary); working with legislators (who she didn't trust); or a ballot initiative, which would allow state residents to vote directly. The third option seemed best. "We need to do this ourselves," she told a coworker who had signed on to the effort. But "this" was a hopelessly daunting task. To even get their initiative on the ballot, they would need to collect more than three hundred thousand signatures—a number well over the entire population of Grand Rapids.

Fahey registered a new nonprofit, Voters Not Politicians. Ballot initiatives were typically led by well-funded organizations who worked with lawyers to craft the language of amendments and raised money before announcing their goals. Voters Not Politicians opened a bank account one day before their first press conference. They didn't even have pens.

As Fahey recalls, they were thoroughly "reamed" by reporters. Local press brutalized their efforts. One article highlighted every time Fahey said the word "like" during the press conference, weaponizing her age, gender, and informal style to paint her as a lightweight. Rumors circulated that she might be a political operative running a fake grassroots campaign. Actual political operatives called, warning that her failure would set back anti-gerrymandering efforts for years.

Depending on the report, Fahey was a naive child, a secret agent, or a destructive amateur. But everyday people reacted differently. Over thirty-three days, Voters Not Politicians held thirty-three town halls across every Michigan district. The more Michiganders learned about partisan gerrymandering, the angrier they became. Fahey then gave people something to do about it. Voters Not Politicians crowdsourced language for their proposal, asking people what *they* thought would be fair. "You could see that people had never been asked before in their life," Fahey remembers, "they were so excited." Creative maladjustment cascaded across the state.

Soon, stops on their tour—to discuss the dry details of state political procedure—were standing room only. Within months of her Facebook post, thousands of volunteers had joined the movement. These were Tea

Party conservatives and Progressives, retirees and students, lawyers and laborers.

In her vanishing spare time, Fahey ran and performed in an improv comedy troupe. A core principle in improv is known as "Yes, and…" A performer accepts whatever their teammate throws at them and builds on it, jokes building unpredictably in the space between them. Voters Not Politicians strikes me as following that ethos. Volunteers came as they were and helped the way they could. A veterinary student researched case law from 4:00 to 6:00 a.m., passing along what she learned to the morning shift. A woodworker carved clipboards for collecting signatures.

Voters Not Politicians wrote up Proposal 2, a ballot initiative that would replace politicians with a commission of citizens—four Democrats, four Republicans, and five Independents—to oversee the districting process. If passed, Proposal 2 would fundamentally change Michigan's political landscape, moving power from back rooms to the voting booth. Their canvassers blitzed the state, spreading the gospel of voter empowerment to people in their homes and on sidewalks across cities, towns, and villages. "They were everywhere," one local reporter marveled. "You couldn't go to an event without seeing them."

Voters Not Politicians pounded the virtual pavement, too. A Facebook counter displayed each signature the campaign logged. The group live streamed milestones. When they reached their final goal, organizers filmed the boxes of signed forms being piled into a U-Haul. It rumbled to the statehouse, where a crowd of cheering supporters had already gathered.

After Proposal 2 had cleared the bar for public support, Voters Not Politicians faced a lawsuit from well-funded opposition, which quickly rose to the Michigan Supreme Court. Of its seven justices, five were Republican—the party benefiting from gerrymandering at the time. Some judges faced intense pressure to side against Voters Not Politicians. Fahey gives off the energy of a bullet train, but the lawsuit deflated even her. Hundreds of thousands of people had raised their hope, and it could all disappear in the hands of half a dozen judges—silencing people in their very effort to be heard. "If it's going to be this rigged," Fahey wondered, "is democracy even worth it?"

But there was no point waiting quietly. Voters Not Politicians organized a presence at the trial. As the case against Prop 2 was argued, the court was filled with citizens and hundreds more waited outside. If the justices were going to take away the people's voice, they'd have to do it in front of them.

The court dismissed the cases, clearing the way for Prop 2 to land on the ballot in 2018. It passed in a landslide with more than 60 percent of the vote, winning across sixty-seven of eighty-three Michigan counties. The following year, the state mailed applications for the redistricting committee to randomly chosen Michiganders, and sixty-two hundred asked to be considered. The final committee finished drawing the state's new districts in 2021.

According to the polling website FiveThirtyEight, Michigan's new districts are some of the least biased in the country, meaning that congressional seats truly represent the majority of votes. Several other states have recently adopted districting committees like Michigan's. Fahey now works as executive director of The People, a nonprofit organization that advocates for voting rights and related issues at the national level.

To me, the details of this story are mind-bending. A woman with no legal experience, barely old enough to rent a car, took on entrenched political interests and won. She shed light on one of the most insidious threats to American democracy, and strengthened the votes of ten million citizens.

Fahey's story is also a challenge because it makes clear just how powerful any of us can be. Her audacious campaign was publicly mocked as naive, but Fahey hopes it will offer a lesson for younger generations. "Kids in Michigan are still going to learn about gerrymandering," she reflects, "but my god, they also get to learn that in our state, that's not what happened." Before her campaign, Fahey didn't think one person could make a difference. She still doesn't. It took thousands. But those people are everywhere. When you teach them about a problem and give them ways to help, they often do.

Growing the Tent

How ambitious can our hope be? Change-makers like Václav Havel and Katie Fahey push toward progress. What about people who stand in its way? Will history bowl them over, or can they come along? These questions have animated Loretta Ross for half a century. Ross has led movements for reproductive rights and racial justice. In Washington, DC, she headed the nation's first rape crisis center and organized the 2004 March for Women's Lives, at that time the largest protest for reproductive rights in US history.

Ross's creative maladjustment is rooted in personal trauma. When she was fourteen, an older family member raped her; she became pregnant and gave birth to a son. After learning she was a parent, Ross's high school refused to reenroll her, relenting only when her mother threatened to sue. Her parents' love and fierce advocacy kept her alive. Ross excelled in advanced STEM classes and by sixteen was enrolled at Washington, DC's Howard University.

In the nation's capital, Ross studied physics and organic chemistry, but spent her spare time protesting race discrimination, the Vietnam War, and South African apartheid. She was tear-gassed before she was old enough to vote. After college, it became clear that activism, not science, would define her life.

Ross focused on organizing, supporting, and uplifting her community, and challenging everyone else. She criticized white women who seemed unaware of the struggles faced by women of color, drawing sharp lines between feminist groups. Then, in the 1970s, an unexpected experience changed her perspective. A letter arrived at the DC Rape Crisis Center from Lorton Reformatory, a prison about twenty miles away. A note inside, written by a man named William Fuller, read "Outside I raped women. Inside I raped men. I'd like not to be a rapist anymore."

"It pissed me off," Ross tells me. "Here we are scraping together resources for rape victims, and now a damn *perpetrator* wants our help?" Her colleagues agreed, and most recommended ignoring Fuller. But she didn't throw the letter away. It sat for months on her desk, "a sore tooth among the pile of papers." Eventually, she decided to visit Lorton, not to help Fuller, but to tell him off. "I couldn't do anything to the people who raped me… but I figured I could make his life more miserable." What awaited her defied expectations. Fuller was not alone, but with several other men, most convicted of sexual assault. The group had gotten their hands on Black feminist literature and were holding discussions about it, but craved guidance from someone doing anti-rape activism. Would Ross help them?

Too stunned to do much else, "all I could do was tell my own story," Ross remembers. "And that opened the floodgates to their stories." Each man had been a perpetrator, and some had also been victims of abuse or molestation. The men named their group Prisoners Against Rape and worked with Ross for three years to educate themselves and become allies. None of this took away what they had done or made Ross their friend. Ten years later, she ran into Fuller in DC. He thanked her for turning his life around and shared that he was now a husband. *Yeah*, she thought, *but you were never supposed to be out on the streets again.*

And yet, the experience changed Ross. Knowing their stories, she could no longer define these men purely by their crimes. They were flawed people who had done terrible things, and who wanted desperately to be part of something better in the years they had left. This became a prequel for Ross's new, more expansive approach to change.

Like her younger self, many activists have little patience for anyone outside their movements. One of their weapons is "calling out": publicly shaming people for inappropriate behavior. Callouts can hold power to account and expose injustice—creative maladjustment made speech. They can also splinter social movements. Change spreads across people like a wave; some wake up to new ideas before others. Ross thinks that when individuals bludgeon people who fail to keep up, they "increase the chances someone will double down and continue" with old patterns of thought and action.

These moral purity tests limit diversity of thought within social movements, freeze out potential allies, and feed what Ross calls "the cannibalistic maw of cancel culture." And like other forms of cynicism, callouts deny people the capacity to change. In Ross's eyes, these fixed views "mirror the prison industrial complex" that most activists abhor.

Ross offers an alternative, "calling in," which she describes as "calling out with love." Calling in names the harm someone does *and* their capacity to grow. Ross used this strategy with Prisoners Against Rape, and in later work deprograming Ku Klux Klan members. "When you ask someone to give up hate," she reasons, "you need to be there for them when they do." Ross relates the story of a personal call-in from which others might learn. A relative of hers would often share bigoted opinions about Mexican and LGBTQ people. One night at dinner she replied to one of his rants, saying, "I know you're a good man. I think you would run into a burning building and rescue somebody if you could and you wouldn't care whether they were gay, straight, Mexican, white. How can I reconcile the good man that I know you are with the words that just came out of your mouth?"

Calling out whittles down social movements; calling in broadens them, creating space for more people to join in. It draws from restorative justice, a practice for moving forward after harm, which also inspired LaJuan White's approach at Syracuse's Lincoln Middle School. Now a professor at Smith College, Ross teaches classes on calling in to new generations of activists. She's clear that no one is obliged to engage in "nonproductive conversations" with provocateurs or trolls, or to put themselves at risk in the service of dialogue. But her new version of social change is fueled by hopeful skepticism, and belief in the goodness of most people.

It's easy to evoke hope by thinking of past victories. But in their own time, the fates of social movements in Alabama, Stonewall, Prague, and South Africa were unknown. Countless movements now churn around the world. Democracy falters, and citizens fight to keep it alive. Authoritarian leaders strip away rights, and people protest and organize to broaden them. It's unclear who will prevail. The boulder rolls down the hill; someone pushes it back up.

Loretta Ross holds no illusions about human nature. "It's not that you don't pay attention to the horrible things people can do," she says, "but you absolutely believe that we can do better." And better has a lot on its side. Across the decades, progress has moved forward. In fits and starts, the boulder has gained altitude. As Ross tells me, "The enemies of human rights think they are fighting us, but they're actually fighting forces way beyond their control. They're fighting truth, evidence, history, and—most of all—time. And truth, evidence, history, and time are going to kick their ass."

Chapter 11

Our Common Fate

On a recent drive to summer camp, my seven-year-old daughter brought up one of her favorite places on earth: her mother's hometown of Tuscaloosa, Alabama. Our recent visit to the South had been an explosion of cousins, backyard sprinkler parties, and sugary snacks. It was perfect, except for the heat. Alabama in July is never comfortable, but this year it had been well over 100 degrees on most of our days there.

"It's probably going to get hotter," I let slip. This did not faze her, because she knows much more about the climate than I did in second grade. She did have questions.

"Will it get so hot that it's on fire?"

"I don't know," I replied, "but it might get harder to live there."

"Will San Francisco be on fire, too?" she pressed on.

"I don't know that, either. I'm sorry to not have better answers," I deflected, knowing that what I am actually sorry about is that the answers to her questions are so awful. In the years to come, she and her sister will awaken to a slow-motion catastrophe they did not cause, one that will shape and perhaps limit their lives.

There are countless spots in our culture where positive momentum is growing. The more I've paid attention to them, the more hope I've felt. But all of this has been dwarfed by my least hopeful stance. Over the last several years, I've become a climate doomer.

There's plenty of reason to feel terror in the face of the climate crisis. Humanity is on track to nearly double the amount of warming agreed upon in the 2015 Paris Climate Accords. In 2022, a billion-dollar natural disaster occurred once every three weeks in the US alone, four times more frequently than in the 1980s. Over 150 million people live on land that will be underwater by 2050. This pain is felt disproportionately by poor countries, who contribute least to the problem.

Around the world, alarms are sounding, and we should be alarmed. Instead, I often feel exhausted by the enormity of the problem. That sort of "doomerism" isn't outrage or fear; it's fatalism—the belief that nothing can save the environment. In a 2019 poll of more than fifty thousand people across many countries, more than half agreed that "climate change is an unstoppable process." A 2021 survey found that young people are two-thirds more likely than older adults to experience climate despair. Who can blame them? Their share of this century is larger, as is their share of its problems.

Doomerism has become the wallpaper of our public imagination: so prevalent that it's hard to notice. So has cynicism about others' climate ideals. In 2022, psychologists asked thousands of Americans how many of their fellow citizens supported strong policies to protect the environment. The average respondent estimated that less than 40 percent did. In other words, most people think that most people don't care very much about the planet. They feel hopeless about the future because they have little faith in one another now.

By now, I hope this makes you skeptical. I hope you're asking: Is it reasonable to write people off like this, and give up on our chances of addressing the climate crisis? As it turns out, doomerism is built on a popular but shaky view about human nature.

A Tragic View of Life

You might not have heard of Garrett Hardin, but he's almost certainly shaped how you see the future. Hardin contracted polio when he was four years old, spending weeks in bed with fevers so high he hallucinated. In school, he was bullied as a "cripple," escaping into violin, theater, and eventually science. Later in life, he developed post-polio syndrome, an agonizing condition that confined him to a wheelchair. He and his wife, Jane, were members of the Hemlock Society, a group that supported people's right to die when they choose. When she developed ALS, the Hardins ended their lives together after sixty-two years of marriage.

Hardin had always seen death as a crucial part of life. His father, Hugh, worked as a traveling salesman, pulling the family from Missouri to Memphis to Chicago alongside him. Beginning at the age of ten, Hardin spent summers in Butler, Missouri, at a family farm he called "the one stable place in my life." Given the limitations brought on by polio, his main chore on the farm was tending about five hundred chickens, and killing one each day for lunch. Meanwhile, he regularly witnessed residents of Kansas City driving their unwanted house cats to the countryside. These city dwellers probably imagined their pets would find a good home, but as the population of strays swelled, feline fever ripped through it. Most cats would die sick, or in the jaws of the farm's fox terrier.

Decades later, Hardin recalled this as a crucial lesson from the farm. Killing was not always cruel; sparing something was not always kind. "All my life," he told an interviewer, "I have been haunted by the realization there simply isn't room for all the life that can be generated." That was true for cats, and even truer for people, whose numbers were exploding around the world.

Hardin studied zoology and biology, eventually teaching at the University of California, Santa Barbara, but the dark realizations from the farm stayed with him. In 1968, he distilled them into a short essay called "The Tragedy of the Commons." It asks readers to imagine a pasture shared by many herdsmen. The field will survive if all of them limit their number of cows. But each is tempted to add another to his herd, and another, until the field is chewed down to dirt, cows and farmers perishing together. Zoom out enough, and the little field in Hardin's essay becomes our planet, the herdsmen our species, and the cows our flights, factories, strip mines, and—most of all—our children and their children.

Hardin used the term "tragedy" not to mean a sad story, but in the original Greek sense: an outcome the hero cannot escape, because it is tied to his fate. According to Hardin, humanity's grand tragedy would come from the clash of two immutable laws. First, the planet could not survive a growing human population. Second, people were too shortsighted to notice this, and too selfish to care. "Ruin is the destination towards which all men rush," Hardin proclaimed, "each pursuing his best interest."

He dedicated himself to preventing that ruin by advocating for policies to limit the human population. In the early 1960s, he gave a speech supporting abortion rights, a taboo subject at the time. For years afterward, women approached him on sidewalks to ask where they could find reproductive care. He and Jane found abortion providers in Mexico and referred hundreds of women to those doctors. Hardin thought of their efforts as "comparable to the Underground Railway of pre–Civil War days."

But over time, his ideas became more fanatical. We were treating humanity like city cats, with a false and poisonous mercy. We needed to think more like chicken farmers. It wasn't enough to give people reproductive choices, he argued, freedoms must be taken away. Hardin supported sterilizing people as a method of population control. In an essay called "Lifeboat Ethics," he argued for ending international aid. Richer nations could save themselves—like passengers on a small craft at sea—but only if they allowed famine to ravage poorer ones. Otherwise, Hardin wrote, "the less provident and less able will multiply at the expense of the abler and more provident." He embraced xenophobia, eugenics, and eventually open racism, saying that the "idea of a multi-ethnic society is a disaster."

Hardin genuinely feared for the planet. He also feared humanity, in ways that were morally backward and factually wrong. Though he predicted the human population would grow more quickly year after year, it has plateaued, with population declining in twenty-five nations as of 2019, even before the pandemic.

Most people thankfully do not share Hardin's prejudice, but many still adopt his thinking. "Tragedy" became an unlikely blockbuster and made him a celebrity. He toured the country, speaking to crowds of hundreds. His hosts in San Francisco couldn't find a large enough hall, so he did multiple events there, like a touring rock band. "Tragedy" is still taught to millions of students. Anyone who cares about the environment can be seduced by its simple logic: Humanity is the problem, and always will be.

Hardin thought overpopulation would destroy the earth, yet he and Jane had four children of their own. This might seem like an odd tension between the couple's fears and their actions. Yet today's doomers are less likely than other people to buy fuel-efficient cars, invest in solar energy, or join rallies for climate justice. If there's no tomorrow, why not indulge today? If everyone else will add another cow to their herd, why bother cutting back yourself?

As we've learned, cynics are surprisingly bad lie detectors—they assume the worst about everyone, and thus have a difficult time telling apart real and imagined culprits. When it comes to the climate crisis, there *are* real culprits. Nearly two-thirds of industrial carbon emissions over the past 150 years originated in just ninety large companies. The top 1 percent of global earners produce twice as much pollution as the entire bottom half.

Elites in the energy industry pull powerful levers to uphold the status quo, for instance, lobbying senators to oppose new energy policies and paying scientists to spread climate disinformation. Cynicism helps their cause. When we decide that *everyone* is too selfish to protect the planet, we give cover to the people and companies harming it most.

Take the idea of a "carbon footprint," the amount of greenhouse gas a person, family, or community emits. If you think about climate change, you probably worry about your footprint. Online calculators can spit out colorful displays of how much you, personally, are harming the environment, and how you can do better—eat less red meat, bike to work, vacation locally.

It's great to make planet-friendly choices. But it turns out that the very idea of "carbon footprints" was invented by British Petroleum, in what one expert calls "one of the most successful, deceptive PR campaigns maybe ever." Over two years, BP invested millions of dollars into reframing the climate crisis. The cause, they argued, was individuals' careless consumption. The cure, they insisted, rested with us as well. "It's time to go on a low carbon diet," BP's ads declared. Energy executives "carbon shamed" activists for flying and eating meat, using callouts to discredit and divide the climate movement.

Our carbon footprints are also intertwined with the structures around us. It would be easier for people to use less energy if there were more electric vehicle charging stations, bike lanes, and clean energy alternatives—options that Big Oil has lobbied against. The carbon footprint is part of a long line of marketing that deflects responsibility for problems away from companies and onto people, all while creating systems that keep people dependent on those companies.

These campaigns wield cynicism like a scalpel. We are all at fault, which is the same as saying no one is *especially* at fault. And while BP guilted people into picking tofu over steak, it did little to change its own massive footprint. In 2018, the company devoted just 2 percent of its budget to renewable energy. The following year, it acquired a new oil patch through its biggest deal in two decades.

Doomerism saps our energy—and like other forms of cynicism, it's built on assumptions about humanity that are just plain wrong. Americans *think* that barely one-third of the nation supports aggressive climate reform. The real number is closer to two-thirds. If you want US policy to preserve the environment, you are part of a supermajority, one you might not realize is all around you.

Most people want a sustainable future, and millions are taking action toward it, driven by creative maladjustment. Many more are already living in sustainable ways, giving the rest of us a recipe to follow.

The Victory of the Commons

In 1976, Hardin took his "Tragedy" tour to Indiana University. Elinor Ostrom, a professor there, sat in the audience and was repulsed, especially by his insistence on sterilizing people after their first child. As she remembers, "People said, 'Well don't you think that's a little severe?' 'No!' [he replied.] 'That's what we should do, or we're sunk.' Well, he, in my mind, became a totalitarian."

She was also skeptical of "Tragedy." "I thought, *He's just made this up*... He said, 'Imagine a pasture open to anyone.' He didn't say, 'Here's my data.'" Hardin invented a fable in which people would greedily overrun the environment, but no one had bothered to do the science. Thirty years later, she would win a Nobel Prize for looking more carefully and revealing the many ways Hardin was wrong.

Ostrom had grown up in Los Angeles during the Great Depression and became a rare working-class student at Beverly Hills' ritzy high school. She loved math but her teachers refused to let her take advanced classes, claiming it'd be no use when she was "barefoot and pregnant." Later, economics PhD programs rejected her—because she had not taken advanced math classes. Ostrom met similar barriers throughout her career and broke through them all, earning her PhD in political science and becoming a renowned researcher.

Her early work focused on a real-world commons problem in her home city. LA depended on water from California's Central and West Basins. These vast underground reserves were plentiful in the early 1900s, but as people, power plants, and resorts crowded into the area, water became scarce. When residents used too much too quickly, seawater flooded in and contaminated the aquifers. By mid-century, overuse was a major problem. Salty water killed grass in parks and on schoolyards. And still, any one person could use as much water as they wanted without facing consequences.

The dilemma could have been pulled straight from Hardin's essay. Yet no tragedy occurred. Instead, citizens self-organized in what Ostrom called "public entrepreneurship." They formed a patchwork of water associations, monitored and regulated usage, and educated the community about the importance of conservation. In the face of a common threat, Angelenos unified. They acted less like *homo economicus*, and more like *homo collaboratus*.

Along with her husband, Vincent, and dozens of students, Ostrom discovered successful commons projects around the world. In Valencia, Spain, which contains stretches of arid but fertile land, farmers must carefully preserve water. For centuries, they have done so with an elaborate turn-taking system, and a "water tribunal," in which cheaters are brought before their peers. Fishermen in Maine self-organize around rules—for instance, throwing back lobsters under a certain weight—using reputation as a carrot and stick. On the mountain meadows of Törbel, Switzerland, farmers share land for growing vegetables and raising cattle. In 1517, they came together to declare that "no citizen could send more cows to the alp than he could feed during the winter." For a half millennium, they have successfully shared the land.

Each group Ostrom studied solved commons problems in its own way, but across all of them she noticed a set of "design principles" for sustainable living. People agree, democratically, about how much of a resource each one of them should use. They elect monitors who can track who is following those rules. Violators are punished, but sanctions are mild, escalating only after repeated offenses. Ostrom found one more thread tying together victorious commons: trust. "If people have rules imposed on them, and don't trust the process...they'll cheat whenever they can," she reflected. When people have faith in one another, they invest more in a future together. Ocean Villages are more sustainable than Lake Towns.

Underlying Ostrom's work was a profound, simple rebuke of cynicism. Hardin asked us to imagine a pasture; Ostrom went out and found real ones. Hardin was blinkered by a view of life in which human greed knew no bounds. But that is just one way people live, especially when they are boxed in by economic and social systems that pit them against one another.

If we search beyond these narrow borders, we find a panorama more beautiful and complex than Hardin ever imagined. Throughout history, people have lived in rhythm with nature—taking what they need and leaving the rest. According to one of Ostrom's students, at least a billion people today sustainably govern themselves and their commons. Millions more live by philosophies such as "Ubuntu," the Bantu notion that "a person is a person through other people." Especially outside of the West, people share their identity with neighbors, ancestors, and descendants.

By comparison, living like *homo economicus* appears lonely, tragic, and optional. Rapacious greed is not hardwired into us. We are not, by birth, the planet's enemies. Caring for ourselves, one another, and the future can all be the same thing.

Choosing a Future

Ostrom studied small communities, where trust and sustainability come naturally. Our most vital commons problem, decreasing global carbon emissions, involves every nation and billions of strangers working together, stretching the idea of community to a breaking point. But people who want to join in this vast effort have more options than ever. The cost of solar and wind energy has plummeted in the last decade, while interest in these technologies has soared. In 2023, the world invested $1.7 trillion in renewable energy, compared to $1 trillion in fossil fuels. Renewables are expected to take over coal as the largest source of electricity in 2025.

These technologies can slow carbon emissions but probably won't get us anywhere near the goals set by the Paris Climate Accords in 2015. "Even if you do all of the electric vehicles, all of the solar, all of the wind," the climate strategist Gabrielle Walker told me, "there's still this crushing gap, which means we can't make it."

I had called Walker to treat my own doomerism, and this comment from her did not seem like a good start. When I asked if she ever feels climate anxiety, she replied, "Every day." But she continued, "I can't afford the luxury of despair. It's too expensive." Hopelessness can stop us from seeking the solutions we desperately need. Walker has remained restless, and what she's learned leaves her with a startling sense of possibility.

As a young girl, Walker took long strolls through the English countryside, keeping detailed field notes on local plants and animals. Recognizing her daughter's passion for nature, Walker's mother sewed an Elizabeth Barrett Browning poem into a tapestry on Gabrielle's wall: "Earth's crammed with heaven / And every common bush afire with God." Walker sees divinity in all forms of wilderness; she revels in dew drops near her home and has taken over a dozen trips to the North and South Poles. Of the endless arctic tundra, she reports simply, "It makes me feel small and I like that."

After earning her PhD in chemistry, Walker joined the magazine *Nature* in 1992 to cover its climate beat. She traveled the world, often touring natural settings scarred by human activity. In a Madagascar rain forest, she watched lemurs galivant and hawks soar, but nearby, trees had been clear-cut, land grazed and left mostly for dead. This "wanton destruction," as Walker calls it, was warping the ecosystem. Primates could no longer produce enough milk, so birth rates fell. Birds used the temperature as a signal to start their migration. As the climate changed, they began their voyage too late, arriving at dried-up lakes. "Instead of a finely tuned orchestra," she tells me, "you just had this cacophony, this mess."

Walker realized, earlier than many of us, that climate change is already here. She focused on solutions, but soon realized that slowing emissions alone could not address the massive climate debt our species had accrued. Driving straight into a wall, we were easing up on the accelerator instead of hitting the brakes.

Walker began a search for additional approaches to the climate crisis. She had heard of carbon removals, technologies that extract pollution from the environment, but they seemed far-fetched and impractical. A trip to Squamish, about an hour north of Vancouver, changed her mind. There, a company called Carbon Engineering operates a massive air capture plant. Their technology pulls carbon dioxide directly from the air, purifies it, and sequesters it deep underground. In 2018, she visited the plant to learn more about its approach and was shocked by how effective it was. *Bloody hell*, Walker thought, *this could work*.

She has since become a fierce advocate of carbon removal techniques. Not all are as high-tech as air capture. The volcanic rock basalt naturally reacts with carbon dioxide, trapping it in solid form. Grinding basalt up and spreading it on the ground speeds the process while improving soil quality. More broadly, carbon removal offers an intoxicating possibility. Instead of slowing the car down, we could turn it around. Instead of mitigating climate change, we could give the earth a chance to heal.

Carbon removal has attracted an explosive amount of investment from companies and governments alike, along with loads of criticism. It is new, untested, and in its current form, too expensive to make a dent in the climate crisis. Some scientists and activists think of it as a rosy distraction from serious climate efforts, or worse—a misdirection oil and gas companies use to pretend they care while continuing their destruction.

"There's a strong fear that even talking about it will cause us to slow down [climate efforts]," Walker says. But in response, she asks: "If you don't like this, what is your plan?" Dismissing carbon removal as unrealistic sounds familiar to her; people said the exact same things about solar and wind power in years past. More importantly, Walker thinks that in the face of our species' greatest struggle, we must try every option. In a world of "ors," she favors "ands."

If renewable energy and carbon removal are two routes to protecting the climate, Walker's third "and" is accountability. Fossil fuel companies could use removal as a distraction, just like they could keep guiding the public toward carbon footprints. But people can refuse to take the bait. Instead, we can vote for regulation and keep an eye on which companies are taking steps to comply with global climate goals. As for corporations who don't, Walker says, we must "come down like hellfire" to compel them.

For years, young people have asked Walker what they can do to protect the environment. She used to talk about changing light bulbs and insulating attics. That's changed, she tells me. "Now when kids ask, I say, 'Make a nuisance of yourself.'"

Not that they need her permission. It's been more than five years since Greta Thunberg decided to skip three weeks of ninth grade and sit in front of Sweden's Parliament holding a sign reading "School Strike for the Climate." Seven months later, more than a million children across 125 countries joined her in a global day of protest. Thunberg's lonely resistance had ballooned into the Fridays for Future movement and galvanized a generation.

Young people are now firmly part of climate leadership, wielding a new sort of political power. They have repurposed social media to build awareness and creative maladjustment—often combining fury, humor, and optimism in viral messages. In 2020, the energy company ConocoPhillips received approval for the Willow Project, a massive, thirty-year-long oil drilling effort on Alaska's North Slope. The oil generated by this project will produce about as much pollution as adding two million gas vehicles to the roads. On TikTok, videos using the hashtag #stopwillow amassed half a billion views. This energy quickly broke social media containment. The White House received over a million letters and thousands of calls to stop the Willow Project.

Willow went through, nonetheless. But behind this apparent defeat are thousands of new activists pushing for the next cause. And though young people are understandably prone to doomerism, many are also taking causes head-on. Philip Aiken, a podcaster who focuses on sustainability, sees hopelessness as a form of privilege. "'It's too late' means 'I don't have to do anything, and the responsibility is off me,'" he says. This sentiment is wrapped up in "Ok Doomer," a sassy reply to climate nihilism that plays on young people's snark toward older "boomers."

The next generation of activists doesn't have time to mourn a planet they must still inhabit. To fight doomerism, they mix news about destructive policies with asset-framed stories about climate wins. Wanjiku Gatheru, founder of Black Girl Environmentalist, regularly shares rundowns of positive environmental news with her thousands of followers. To Gatheru, these are not just feel-good messages, but fuel for further work. "Fear doesn't motivate people towards sustainable action," she explains. "Providing solutions in the midst of discussion of a problem helps get people engaged."

As younger people age into voting, they will bring environmental issues further into the center of political life. In the 2020 US election, nearly a third of voters under the age of thirty named climate change as one of their top three issues. Civic pressure on leaders will rise steadily in the years to come and is already driving landmark decisions at every level. The 2022 Inflation Reduction Act included the most domestic support of climate action in US history. That same year, at the United Nations Climate Change Conference, more prosperous nations agreed to "compensation pledges" to support poorer nations that suffer most from the climate crisis. In 2023, a judge ruled in favor of plaintiffs ranging from five to twenty-two years old who sued Montana for failing to provide them a "clean and healthful environment." Similar suits are moving forward in Hawaii, Utah, and Virginia. And just weeks after the Montana decision, the Biden administration barred oil drilling on more than ten million acres in Alaska, canceling leases previously given to oil companies.

A few weeks after my daughter and I talked about a world on fire, we took part in a beach cleanup along the Pacific Coast. We had joined cleanups in our neighborhood before, but it was different to congregate by the sea's wild expanse—to be close to the world we want to protect, side by side with other people on the same mission. A huge number of volunteers were young children. It struck me that as this generation wakes up to the crisis, many will feel furious, terrified, and defeated, as is their right. But they will also fight—some through technologies that surpass our imagination, others in old-fashioned ways, by protesting, leading, legislating, or just cleaning up.

The world's greatest disaster has also generated a historic global movement, one that could drive changes that seem like fantasies now, just like the Progressive movement did more than a century ago. This will be led by young people's anger and their efficacy. But too often, they are left to shoulder this movement alone. Wanjiku Gatheru is tired of people telling her how inspirational she is. "Hope is something that you earn. We got our hope because of the hard work we put in every single day," she says, urging people to take action themselves. "Don't borrow our hope. Hope with us."

The point of all this is not to convince you that the climate crisis will subside. That would be optimistic, and the last decades have given little reason for optimism. The point is that we don't know what will happen, and can still make choices that matter. And if we were to find a way to live in greater harmony with the world, that would not be a shocking departure from human nature, but an expression of our deepest values.

The same is true of the many edges on which we now teeter. Will democracy erode further, or battle back? Will people feel more divided year after year, or rediscover shared purpose? Will wealth continue to disappear into the hands of ultra-elites, or can standards of living improve for everyone else? Will mental illness continue to rise, or fall as we rediscover human connection?

I don't know the answer to any of these questions, and neither do you. But in each case, the possibility of positive change lives inside us.

Billions of us are taught that life is a battle where winners take all and losers are everywhere. Our trust and hope are mocked as naive. By now, you know better. You can stay skeptical and avoid knee-jerk assumptions about people. You can remember that the media warps our view of one another and look for better data. Most of all, you can realize that hope is not a weakness, but a path to being less wrong and more effective.

We can use that hope, like a divining rod, to locate others who want the same things we do, building solidarity and common cause. At a climate protest, an audience member asked the writer and activist Bill McKibben what an individual can do to fight climate change. McKibben answered, "Stop being an individual." When Emile brought scientists and peacemakers together after his first surgery, he urged us to teach people about the sublime potential they all have—to walk through darkness and spread light. "And the nice thing is," he said, "that this force is in us and communal. It's not owned. And the best way to activate a communal force is to be a community. That's why we're here."

You were right, Emile. That's why we're all here.

Epilogue

By 2011, I had seen Emile at conferences in Washington, DC; Chicago; and San Francisco—always among larger groups. But we lived just a few miles apart, in Cambridge, Massachusetts. That April, we decided to get together at a café near Harvard Square, the first of a few hangouts before I left for Stanford the following year.

I imagine he was wearing one of his signature flannel shirts. I imagine the café was crowded (it always was) and we retreated to one of its cramped upstairs tables. I must imagine these things, because it's impossible to claw back every detail of those moments and return to a past in which Emile is still alive. Here's what I do remember: He'd just come back from his honeymoon; my wedding was four months away. He and Stephanie had just received a new shipment of bees, and he was lit up with details about the hive. The future felt long and clear.

And here's what I will never forget: Emile believed. We talked about a project we might do together, scanning people's brains while they watched political rivals tell their stories. To me, this was an interesting idea. To Emile, it was a doorway. He *knew*—not in his mind but somewhere deeper—that almost no one is born to hate; that kindness, cooperation, and care are a return to who we really are. He knew that science could give us signposts on the way home.

I remember bouncing from inspired to incredulous and back again, as I always did with Emile. It would be years before I'd understand the pain in which his hope was forged. It would be nearly a decade before life would give him a final test, and his strength would burst through to meet it.

This future is nothing like the one we imagined. As of this writing, it's been just over three years since Emile died. Stephanie tells me the family is thriving. Clara, eleven, has grown thoughtful and introspective in ways that remind Stephanie of Emile. Atticus, now nine, loves building things out of wood like his dad. The two attend a progressive school that shares many values with Peninsula, the community that shaped Emile's early life. Clara will graduate this year and plans to sew a dress for the ceremony from one of his old shirts.

Stephanie's new job at an arboretum brings her surprise and delight each day. She has always loved interacting with the natural world, especially through beekeeping. In 2017, she wrote a book on the subject. When Emile got sick, she let go of this passion, thinking she might never return. The week after he died, her publisher got in touch to see if she wanted to write a second book about bees. It was a balm at the exact moment she needed one. As she remembers, "It was like, here, put energy into this thing that will help you feel like yourself."

Stephanie and the children have grown in new directions. They camp in places Emile never saw and live lives he could not have anticipated. But his memory is everywhere. On crisp fall nights, they sleep in the backyard tree house he built. On their travels, they collect rocks and shells to decorate his grave site. During conversations, they wonder often about what he might have said. He remains a co-parent to Stephanie. In difficult moments, she sometimes screams out, "Where are you?!" But she finds solace in how Emile believed in her. "He taught me to trust my own intuitions...Inside of me is us." Emile is part of their sadness and their strength. "Grief is a muscle," Stephanie reflects. "The weight of what you've lost doesn't diminish but you get better at carrying it."

The last time I spoke with Emile, this is what he most wanted: for his presence from beyond this life to bring his family warmth and comfort. As he wrote to Stephanie, "What I am to you is really a reflection of your own mind." We all, eventually, become someone else's memory, our legacy a series of imprints in a series of lives. For Stephanie, Clara, and Atticus, Emile remains as a tender sort of haunting.

For me, he lives on as a challenge. Emile had a sharp antenna for suffering and possibility. Sitting with him, you could feel both—along with the weight of injustice, the urgency of action, and the goodness of most people. I yearn to relive moments with him, in part because I miss his version of the world.

But I don't need to remember what he said or how he smiled ten years ago. Through the people in his life, his writing, and his science, Emile's philosophy has materialized inside my mind. I've combed through hundreds of studies revealing that hope is a precise, powerful strategy for wellness, harmony, and social change. Even in their dry data tables, I witness Emile.

In my dark moments—which are plentiful—he asks more of me. I realize now that cynicism has become my default mode. For so long, I kept it meticulously hidden, which only let it fester. But more recently, I've treated my inner life like an experiment. Many of my worst assumptions about people have fallen apart under scrutiny. Much of the data I've collected has come back more positive than I imagined. This hasn't turned me hopeful overnight. But new habits of mind and action are taking root. Trust is growing more natural. Openness easier. When new challenges arise or hopelessness creeps in, I ask myself, "What would Emile do?"

One of his favorite authors, Ralph Waldo Emerson, wrote that "in every work of genius we recognize our own rejected thoughts; they come back to us with a certain alienated majesty." In the past, I thought that Emile was hopeful and I was cynical. But looking deeper, I realize that trust, vulnerability, and faith in people have always been inside me. Hope has never been foreign to me, just forgotten. Cultivating it doesn't require me to invent anything, but merely to remember.

Stephanie, Clara, and Atticus talk about ripples: the force each of us exerts, felt in small ways at great distances. Emile's ripples are everywhere. For the hundreds of people he inspired, he remains as a beacon. For scientists, his knowledge is a foundation on which to build. His perspective lives on among educators, leaders, activists, and parents all over the world—even if they'll never hear his name.

Emile is still pulling me out of cynicism and back into myself, though he will never know that. I hope, by now, he might have done the same for you. If so, consider reaching back, finding someone in your life who's run out of hope, and bringing them with you. Use your ripples wisely.

Acknowledgments

It was strange to begin a project on hope and trust by admitting that I often lack both. But the very process of turning those reflections into this project connected me with many people who gave me reasons to believe. Over the past two years, a community of colleagues, friends, and family have made this book possible.

My first opportunity to publicly share my interest in cynicism came at the TED Conference in 2021. The TED team, including Chris Anderson, Cloe Sasha, and Briar Goldberg, graciously gave me this opportunity and were ever present in their support as I prepared. Friends helped, too. Thanks to Dan Gilbert, Angela Duckworth, Liz Dunn, Yotam Heineberg, Amanda Palmer, and Kelly McGonigal for incisive feedback on early versions of the talk.

As I built out the ideas in *Hope for Cynics*, my longtime friend and literary agent, Seth Fishman, provided guidance and encouragement, and when the time came, found a home for the book. I couldn't be happier that home turned out to be Grand Central Publishing. The first time Colin Dickerman and I spoke, his vision for this book sprang off the Zoom screen. His conviction—that it could not only teach but also help people—has guided me since, as has his steady editorial hand. Karyn Marcus, also at Grand Central, challenged me to put more of myself into the pages, making the book more authentic and personal. Ian Dorset provided timely and helpful support through the publication process.

Outside of Grand Central, several professionals also guided *Hope for*

Cynics. Toby Lester provided insightful comments on an early draft. Andrew Biondo lent his impressive knowledge of the ancient Cynics to chapter 1, and Alan Teo his knowledge of *hikikomori* to chapter 6. Evan Nesterak's meticulous fact-checking calmed my neuroticism (and since he has not fact-checked this acknowledgments section, I'm sure there's a name misspelled here somewhere!).

Two Kates went above and beyond their roles in supporting this work. Kate Petrova served as the "scientific auditor" for claims made in this book and cowrote its "Evaluating the Evidence" appendix. She also became a first reader of many rough pages, pointing me to research and ideas I would not have encountered otherwise. Kate Busatto started out as a "story doctor" for this project, helping me locate and finesse some of the narratives in it. But she did much more, diagnosing and healing issues throughout the entire book during the lonely middle of the writing process. I am certain this project would not be what it is without either Kate.

The center of my professional life is running the Stanford Social Neuroscience Lab, and every day I feel grateful to work with this small community of curious, warm, and brilliant scientists. "Snails" supported the writing of this book, many providing helpful comments on early drafts. I especially want to thank students and trainees whose work is featured here: Sam Grayson (gossip and cynicism); Eric Neumann (self-fulfilling trust mindsets); Rui Pei and the "intervention team" (the two Stanfords); and Luiza Santos (cross-party political conversations). Several colleagues from outside Stanford read or discussed early drafts and pages of this work as well, including Adam Grant, Laurie Santos, Mina Cikara, Nour Kteily, and Adam Waytz.

Many people sat for interviews and allowed me to include pieces of their lives in these pages. Thanks especially to "Megan," Andreas Leibbrandt, David Bornstein, Robin Dreeke, Atsushi Watanabe, LaJuan White, Andrés Casas, William Goodwin, Katie Fahey, Loretta Ross, and Gabrielle Walker.

Three friends, collectively known as the "ools," have been my go-to sources for advice and reflection over the last quarter century. Thanks to Eric Finkelstein and Daniel Wohl. Luke Kennedy, the closest I've ever had to a brother, has shaped my thinking more than any academic, and I'm eternally grateful for his presence in my life.

Ethan Kross and I have been friends and colleagues for twenty years, and more recently we both embarked on parallel journeys into science writing. He has since been a stalwart "chatter buddy," trading advice, guidance, and (lots of) reassurance through our many spells of anxiety. It's a joy to support him and a tonic to receive his support in return.

Any book reflects the support of its author's family. My wife, Landon, gave me the room to write and think amid our wall-to-wall busy lives. Our children, Alma and Luisa, are my single greatest reason to keep practicing hope. This book is dedicated to them.

Emile Bruneau is, of course, its north star. Many people shared memories of him with me for this project, a tender and beautiful act full of laughs, tears, and gratitude for this wonderful man, which allowed him to come into focus in my mind. Thank you to Jeff Freund, Janet Lewis, Heather and Tim McLeod, Samantha Moore-Berg, Franck Boivert, Andromeda Garcelon, Emily Falk, Nour Kteily, Mina Cikara, and the staff of the Peninsula School.

Stephanie Bruneau changed this entire book in an instant. During our first conversation, I asked for her blessing to write *at all* about Emile. After a while, I mentioned my regret that Emile and I didn't have a chance to collaborate more deeply. She said, simply, "This can be your collaboration." That opened the door, with her help, to connect with his friends and family and better understand his life so I could share it here. Allowing me to include her and her family's lives in these pages was a titanic act of generosity. Her comment also exhibited a deep emotional wisdom that has shone through in every other conversation we've had. Stephanie recognizes, fundamentally, that connections need not end along with our natural lives. Her strength astounds me.

Emile, I wish we had been closer when you were alive. I wish you had gotten to write your book. Thank you for all your help writing mine, and for lending me even an ounce of your courage and, yes, your hope.

Appendix A

A Practical Guide to Hopeful Skepticism

Hope for Cynics describes the many ways cynicism traps us, and how we can escape its grip. But you might be craving ways to implement this knowledge and practice hopeful skepticism yourself.

This appendix is meant to help you do that. Each of the following exercises provides "treatments" against cynicism based in behavioral science research. Many build on the experiments I tried myself throughout the book. Each can be done in just a few minutes, on your own or with someone else. These are not meant to be grand gestures that change anyone's life in an instant. That's usually not how change happens. It occurs through the new habits we cultivate that redirect us, slowly, toward who we want to be.

As we've learned, cynicism often comes down to mistaken negative assumptions about people. Hopeful skepticism, then, is a matter of opening ourselves to data. The monk and author Pema Chödrön writes, "We can approach our lives like an experiment. In the next moment, in the next hour, we could choose to stop, to slow down, to be still for a few seconds. We could experiment with interrupting the usual chain reaction."

In these exercises, I invite you to approach your life more like an experiment. If you're like most people, the results might surprise you.

Transform Cynicism to Skepticism

When someone lets us down, it's natural to feel disappointed. It's also natural—but less helpful—to become "pre-disappointed," assuming most or all people are just out for themselves. Pre-disappointment drives our feelings and actions and builds cynicism. Yet we rarely put it to the test. Instead, try *skepticism*: a more scientific kind of thinking. Here are some steps for starting to replace cynicism with skepticism.

Connect with your core values. Spend some time reflecting on what you value most in life: Connection to others? Creativity? Intellectual pursuits? Write down why those values matter to you and how you try to express them in your life. These "values-affirmations" can help people remain open to new ideas.

Focus on a safe home base. Cynicism arises quickly when we feel threatened or alone. By contrast, secure and communal relationships give us space to explore what we believe and why. When questioning cynical assumptions, it can help to first ground yourself in those relationships. If you can, think about one or two people in your life whom you trust deeply. Write down what they mean to you, and how you feel around them.

Be skeptical of your cynicism. Pick a cynical thought you have about someone else, people in general, or the world. What information is this belief based on? Especially if it's a general belief, such as "Most people are selfish," ask yourself whether the evidence you have really supports that claim. If not, what evidence would you need to be sure? If you don't have that evidence, the next section will help you get it.

Collect New Social Data

Human beings need to protect ourselves from threats, so it makes sense that we remain vigilant to signs someone might try to harm us. But this can also bias us toward the negative. It's easier to pay attention to, remember, and judge people based on our worst interactions with them, while many positive moments slip out of our consciousness. These exercises will help you take a more evenhanded approach to the social world.

Fact-check a cynical theory. Consider one of the cynical theories you have about people. Now turn that theory into a *hypothesis*: a concrete prediction about how people should behave in a certain situation. Perhaps you think your coworkers are selfish. Now, test that hypothesis directly. For instance, you might ask a small favor of three people you work with. Cynicism might predict that none will step up. If that turns out to be true, you have more evidence for your cynical hypothesis. But if even one of them helps, you might reconsider.

Encounter counting. Think about all the interactions you have on a given day, with friends, family, colleagues, and anyone else. How positive would you guess your average conversation is, on a scale of 1 (very negative) to 10 (very positive)? Write down your answer. Now, instead of guessing, collect the data. Keep a notebook with you one day, and after each conversation you have, write down on that same scale how positive it *actually* was. Compare the real data to your predictions.

Test the social waters. To level up your encounter counting, try predicting and testing not just regular interactions, but also new ones. Think of something you have been meaning to say to a loved one—disclosing a struggle, asking a favor, expressing gratitude—but have been hesitant about. Alternatively, imagine striking up a conversation with a stranger on your way to school or work tomorrow. In either case, predict how positive (1–10) you think this interaction would be, then jump into the social waters and try it. Compare reality to your predictions.

Balance your media diet. Many of us feel overwhelmed and cynical when reading, watching, or scrolling through the news. But many of the stories we encounter emphasize the negative. Try balancing your diet by using outlets that specialize in "asset-framed" stories about positive developments. The next time you read or watch a story that makes you feel hopeless, visit the Solutions Story Tracker (www.solutionsjournalism.org/storytracker) and search for coverage on the same theme that points to positive developments.

Put Hopeful Skepticism into Action

Now that you've rethought your assumptions and collected new data, try giving other people reasons to hope and chances to connect. As we've seen, many actions become self-fulfilling prophecies: People become who we expect them to be and follow our lead. Use that power to create positive ripples where you can.

Use a reciprocity mindset. The next time you are considering trusting someone, remember that your decision will affect not just you, but also them: changing how they feel and what they do in response. Ask yourself, "What positive influence could I have on this person if I gave them a chance to prove themselves? How could my faith in them be a gift?"

Trust loudly. Provided you feel safe doing so, take a leap of faith on someone. This doesn't have to be an enormous act of trust. You can start small: letting your child make some additional decisions on their own, trusting someone you work with rather than looking over their shoulder, confiding a small secret in a new friend. When you do, trust *loudly*, telling that person directly that you're acting this way because you believe in them. Notice what they do in response, how you feel, and how you might use this practice in other situations.

Savor goodness with others. When we gossip—talking about people and their actions—we tend to highlight the negative. Try balancing this instinct by doing the opposite. The next time you witness someone acting kindly, try some positive gossip, "catching" people's best and sharing it in conversations. Notice the helpers all around you, and help other people notice them as well.

Disagree better. If trust is hard to come by in regular conversations, it can feel impossible during disagreements. But at least some conflicts grow toxic because we assume the worst about people on the other side instead of being curious. The next time you find yourself in a disagreement, try to be more precise. If your conversation partner is up for it, ask them not just what they think, but *how* they came to have their opinion: Get to know the story behind it. Next, try to map out exactly what you *agree* on, seeing if there might be more common ground than you think.

Appendix B

Evaluating the Evidence

This book has provided a tour of cynicism, trust, skepticism, and hope, drawing primarily on research in psychology. But, like other sciences, my field is not a collection of immutable facts; it's a dynamic process of continuous refinement and correction.

In recent years, high-profile findings in psychology, biology, economics, and other disciplines have been found to be less robust than scientists initially believed. This evolving landscape has led psychologists to embrace a new ethos of transparency, ensuring that the foundation upon which their claims are built is clearly illuminated for all to see.

I want you, as a reader, to have a chance to understand the evidence underlying this book if you want. In that spirit, I teamed up with a colleague, Kate Petrova, to write this appendix. The two of us identified a list of key scientific claims from each chapter. Kate then did a deep, independent dive into the research on that claim—reading the studies I cite in the book and many more. Based on her research, she assigned a score between 1 and 5 to each claim. A rating of 1 indicates that the claim lacks substantial supporting evidence at this stage. A rating of 5 indicates the claim has strong and consistent support.

We then both reviewed each claim and the evidence behind it to ensure that we agreed on the ratings. If a claim lacked strong evidence, I removed it or changed language in the book to make it clear that more work needs to be done on that subject.

Here are the guidelines we used when assigning a "strength of evidence" number to each claim:

5. **Very Strong Evidence:** A rating of 5 signifies that the claim is firmly established and widely accepted within the scientific community. These claims are supported by many robust, replicated studies and meta-analyses—a type of study that brings together data from many independent projects—and there is a strong consensus among experts regarding their validity.
4. **Strong Evidence:** Claims with a rating of 4 are supported by a substantial body of evidence. Many studies have found consistent patterns across various settings. The reason some of the claims in this category fell short of receiving a rating of 5 is because of a lack of comprehensive meta-analyses or there are minor inconsistencies in results reported across studies.
3. **Moderate Evidence:** This rating usually indicates that there is a growing body of evidence supporting the claim. Findings across multiple studies are relatively consistent, with occasional mixed results. Most of the claims that received a rating of 3 are based on research that is still relatively new. Additional studies, replications, and extensions of this research across a broader range of populations or contexts will help scientists better understand the phenomena in question in the years to come.
2. **Limited Evidence:** Claims in this category derive from only a handful of studies, often with small or nonrepresentative samples. Results might be mixed or have yet to be replicated. Readers should treat these claims with caution and remain open to the possibility that new data could evolve scientists' thinking about them.
1. **Weak Evidence:** A rating of 1 suggests that the claim lacks strong supporting evidence. Claims in this category may be based on just one or two studies, or on a study that was not replicated. Oftentimes,

claims with this rating are based on brand-new research, exciting but still uncertain. Readers should approach claims with a rating of 1 with caution, as confirmation will be necessary to consider them robust.

In the rest of this appendix, you will find a list of the claims the two of us agreed on, along with their ratings. If that claim received a rating of 3 or lower, we will also provide a brief explanation as to why. If you want to dig further into the data, we have included a spreadsheet that contains the extensive research used to evaluate all claims on *Hope for Cynics'* website.

We hope this appendix will give you a chance to better understand the science underlying this book.

Jamil Zaki & Kate Petrova

Chapter Claim Ratings

Introduction:

Claim 0.1: Trust has been on the decline worldwide.
RATING: **5**

Claim 0.2: Kindness and generosity increased during the pandemic.
RATING: **4**

Chapter 1:

Claim 1.1: Cynicism is linked to lower trust.
RATING: **5**

Claim 1.2: Social support softens the effects of stress on individuals.
RATING: **5**

Claim 1.3: Cynicism is associated with poorer physical and mental health, a link that can't be explained by factors like gender, race, or income.
RATING: **4**

Claim 1.4: Trust is associated with group-level positive outcomes and resilience to adversity.
RATING: **4**

Chapter 2:

Claim 2.1: Cynics do less well on cognitive tests and have a harder time than non-cynics spotting cheaters and liars.
RATING: **3**

Relatively few studies have directly tested the associations between cynicism and cognitive abilities. More studies have examined the links between cynicism and social thinking, and generally find that cynical people tend to overgeneralize and wrongly categorize honest individuals as cheaters or liars, leading to problematic social outcomes, such as isolation and breakdown of relationships. Additional research is required to replicate and extend work in this area.

Claim 2.2: Attachment can change with new experiences after childhood.
RATING: 3

Early work on attachment was characterized by a relatively fixed view, suggesting that once an attachment style is formed in early childhood, it remains stable across the life span. More recently, evidence began to accumulate suggesting that attachment style can change as a result of therapy, and that people's attachment security often varies from one relationship to another. This direction of research is still relatively new, and more research is needed before we can confidently estimate exactly how malleable adult attachment is and under what circumstances.

Claim 2.3: Experiences of betrayal can lead to mistrust.
RATING: 5

Claim 2.4: Adversity can have positive effects.
RATING: 5

Chapter 3:

Claim 3.1: Cynicism is hereditary, with less than half of it explained by genetic factors.
RATING: 3

The heritability of cynicism has been investigated in several twin and family studies. While most studies find evidence pointing to a genetic component to cynicism, some studies report none at all. Studies that do find evidence for heritability of cynicism produce varying estimates of the size of this effect relative to environment. The nature of the interaction between genetic and nongenetic factors that contribute to cynicism is complex and not yet thoroughly understood, which is why additional research is needed.

Claim 3.2: Exposure to marketized living increases selfishness.
RATING: 3

Studies from across the world show that exposure to free marketplaces incentivizes individuals to cooperate; however, the motivation behind such kindness remains less well understood. On one hand, marketplaces can incentivize individuals to behave in a genuinely prosocial manner. On the other hand, markets can foster a view of people as self-interested, potentially undermining communal forms of cooperation and increasing cynicism. Several studies provide evidence for both sides of the argument, raising the possibility that the effects of marketized living might depend on a person's context. Additional research is required to more clearly delineate when exchange norms cooperation versus selfishness.

Claim 3.3: Acts of kindness boost well-being.
RATING: 5

Chapter 4:

Claim 4.1: People have a natural inclination to focus on the negative.
RATING: 5

Claim 4.2: Gossip can have positive effects on communities.
RATING: 4

Claim 4.3: People often believe that human beings are good at their core.
RATING: 3

Several studies find that people hold an optimistic belief in the inherent goodness of others. This belief seems to stem from a psychological tendency to view people as having an essential "true self" that is morally virtuous. Though there is initial evidence that this belief is fairly robust across individuals and cultures, this is still a new area of research.

Claim 4.4: People who consume more news media have more cynical views of the people around them.

RATING: 2

The research on this topic is mixed. Some studies find a link between news media consumption and cynicism, while others find none. A concept known as the "mean world syndrome" suggests that frequent exposure to negative news can foster cynicism. Some studies show that viewing satirical news programs in particular is related to cynicism. Exposure to strategic news frames has also been shown to increase issue-specific political cynicism. However, news consumption has also been positively associated with political knowledge and unrelated to cynicism toward politicians in other studies. In sum, though there might be a connection between consuming certain types of news media and cynicism, more research is needed to draw definitive conclusions.

Chapter 5:

Claim 5.1: People underestimate others' helpfulness and kindness.

RATING: 3

Evidence is beginning to accumulate suggesting that individuals often underestimate others' helpfulness and kindness. However, relatively few studies have directly examined this phenomenon. Furthermore, mismatches between people's expectations and reality could be driven by misperceptions of others' kindness or overestimation of one's own altruistic traits. There is also evidence that both actual and perceived helpfulness varies across cultures and situations.

Claim 5.2: Higher expectations lead to an increase in performance.

RATING: 5

Claim 5.3: Trust is self-fulfilling: When we trust others, they are more likely to act in trustworthy ways.

RATING: 5

Chapter 6:

Claim 6.1: Loneliness rates are increasing.

RATING: 3

Some studies report an increase in the prevalence and intensity of loneliness among adolescents and young adults. In contrast, studies of older adults tend to point to more-stable levels of loneliness across time, with some studies even reporting reductions in feelings of loneliness in old age. Large-scale longitudinal studies that track the same individuals across long periods of time are needed, as well as examinations of changing loneliness across culture.

Claim 6.2: Loneliness has deleterious effects on both physical and mental health.

RATING: 5

Claim 6.3: People frequently underestimate how much they would enjoy interacting with others.

RATING: 4

Chapter 7:

Claim 7.1: People are less rational, more kind, and more principled than *homo economicus*.

RATING: 5

Claim 7.2: Organizational cynicism stagnates progress.

RATING: 4

Claim 7.3: Groups that collaborate internally succeed more than those that compete among themselves.

RATING: 5

Chapter 8:

Claim 8.1: People overestimate political out-groups' negative qualities.
RATING: 4

Claim 8.2: Negative emotions are amplified on social media.
RATING: 3

Several large-scale studies show that negative emotions become amplified and spread rapidly on social media. This amplification can have damaging social effects, but more work is needed. The study of emotion on social media is still in its infancy, and additional studies are needed to replicate and extend existing findings across a broader range of social media platforms, users, and contexts.

Claim 8.3: Contact with members of the out-group help build empathy and hope.
RATING: 4

Claim 8.4: Correcting misperceptions about out-group members is an effective way of softening conflict.
RATING: 3

There is initial evidence that correcting inaccurate beliefs about opposing political groups can decrease intergroup conflict and support for undemocratic practices. However, several studies also show that interventions aimed at correcting such misperceptions can also fail to change people's thinking. Additional research with large and diverse samples of participants is needed to better understand when and how misperceptions can be corrected effectively.

Chapter 9:

Claim 9.1: Interpersonal trust predicts support for social welfare policies.
RATING: 2

Only a few studies have empirically examined the links between interpersonal trust and support for social welfare policies. Even though existing studies converge on the finding that more trusting individuals (and societies as a whole) tend to be more supportive of welfare policies, additional research is needed to clarify the magnitude of such associations, as well as to establish the extent to which this pattern generalizes across social, cultural, and political contexts.

Claim 9.2: Scarcity decreases mental capacities.

RATING: 2

The research on this topic is mixed, with some studies finding evidence that experiences of scarcity and poverty reduce cognitive resources, and others failing to find such effects or finding only weak evidence. Two studies conducted in India provide experimental evidence for the cognitive impacts of scarcity. However, subsequent reanalyses seem to indicate that the original findings of at least one of these studies lack sufficient robustness. Furthermore, similar studies failed to replicate these patterns of results in a Tanzanian population.

Claim 9.3: Direct cash transfers are an effective way to lift people out of poverty.

RATING: 4

Claim 9.4: Direct cash transfers are not associated with undesirable outcomes like system abuse or increased spending on "temptation goods."

RATING: 5

Chapter 10:

Claim 10.1: Cynical people are less likely to challenge the status quo (e.g., by engaging in activism) on issues that matter to them.

RATING: 4

Claim 10.2: When a committed minority reaches a size of roughly 25 percent of the total population, there is a significant increase in the adoption of a new social norm.

RATING: 3

Several studies show that people's willingness to participate in social movements or to adopt new norms depends on how many other individuals they perceive to be participating. One study found a "tipping point," such that when a committed minority reaches a size of roughly 25 percent of the total population, new social norms are more likely to spread. Despite this initial support, relatively few studies have empirically examined this question, leaving unclear how much this phenomenon generalizes across different types of cultures and situations. Other studies indicate that individuals' perceptions of support for a movement within their close personal network matter more than their perceptions of how many people have joined a cause.

Claim 10.3: A sense of self-efficacy can motivate people to fight against injustice.

RATING: 5

Chapter 11:

Claim 11.1: Americans underestimate their fellow citizens' support for climate reform.

RATING: 3

Only a few studies to date have examined Americans' perceptions of others' climate attitudes. These studies show that people incorrectly believe there is little agreement on the existence and severity of climate change and less support for climate reform than there actually is. At least one study documents these misperceptions in policymakers as well.

Claim 11.2: Hope motivates climate action.

RATING: 5

Claim 11.3: When faced with the problem of the commons in real life, people often find ways to cooperate instead of acting selfishly.

RATING: 4

Claim 11.4: When individuals' sense of self extends beyond their immediate surroundings and life span, they are more likely to cooperate and contribute to conservation efforts.

RATING: 4

Notes

Introduction

My lab and I have discovered: Karina Schumann, Jamil Zaki, and Carol S. Dweck, "Addressing the Empathy Deficit: Beliefs About the Malleability of Empathy Predict Effortful Responses When Empathy Is Challenging," *Journal of Personality and Social Psychology* 107, no. 3 (2014): 475–93; Sylvia A. Morelli et al., "Emotional and Instrumental Support Provision Interact to Predict Well-Being," *Emotion* 15, no. 4 (2015): 484–93.

"Even when she was in the pits of despair, she only gave me light": This quote, like many in this book, is drawn from materials shared by Emile's family and friends. In this case, credit goes to Shawn Kornhauser, who recorded a talk that Emile gave shortly after his diagnosis; see Annenberg School for Communication, "Emile: The Mission of Emile Bruneau of the Peace and Conflict Neuroscience Lab," YouTube, October 9, 2020. Other contributors to Emile's presence in this book can be found here in the notes, and in the acknowledgments.

Emile died on September 30, 2020: Jeneen Interlandi, "The World Lost Emile Bruneau When We Needed Him Most," *New York Times*, November 2, 2020.

Racial tensions ran high: White House Historical Association, "Racial Tension in the 1970s," accessed October 11, 2023, https://www.whitehousehistory.org/racial-tension-in-the-1970s.

By 2018, only 33 percent felt that way: One possibility is that this change reflects new demographics joining the GSS. For instance, white, middle-class men have more reasons to trust the American economic system than historically marginalized groups. And indeed, some groups (for instance, Black Americans) are less likely to report high trust on the GSS. However, my team's own analysis of the GSS data find that trust declined even when controlling for race, gender, age, and income. This also matches with a 2011 analysis finding that much of the trust decline is driven not by the addition of marginalized populations to the GSS, but rather a decline in trust among white respondents. See Rima Wilkes, "Re-Thinking the Decline in Trust: A Comparison of Black and White Americans," *Social Science Research* 40, no. 6 (2011): 1596–1610.

tendency is to distrust: Edelman, "2022 Edelman Trust Barometer," accessed October 11, 2023, https://www.edelman.com/trust/2022-trust-barometer.

Between the 1970s and 2022: Gallup, "Confidence in Institutions."

cynics suffer more: B. Kent Houston and Christine R. Vavak, "Cynical Hostility: Developmental Factors, Psychosocial Correlates, and Health Behaviors," *Health Psychology* 10 (1991): 9–17; Susan A. Everson et al., "Hostility and Increased Risk of Mortality and Acute Myocardial Infarction: The Mediating Role of Behavioral Risk Factors," *American Journal of Epidemiology* 146, no. 2 (1997): 142–52; Tarja Heponiemi et al., "The Longitudinal Effects of Social Support and Hostility on Depressive Tendencies," *Social Science & Medicine* 63, no. 5 (2006): 1374–82; Ilene C. Siegler et al., "Patterns of Change in Hostility from College to Midlife in the UNC Alumni Heart Study Predict High-Risk Status," *Psychosomatic Medicine* 65, no. 5 (2003): 738–45; Olga Stavrova and Daniel Ehlebracht, "Cynical Beliefs About Human Nature and Income: Longitudinal and Cross-Cultural Analyses," *Journal of Personality and Social Psychology* 110, no. 1 (2016): 116–32.

harder time spotting liars: Nancy L. Carter and Jutta Weber, "Not Pollyannas," *Social Psychological and Personality Science* 1, no. 3 (2010): 274–79; Toshio Yamagishi, Masako Kikuchi, and Motoko Kosugi, "Trust, Gullibility, and Social Intelligence," *Asian Journal of Social Psychology* 2, no. 1 (1999): 145–61.

less likely to take chances again: Victoria Bell et al., "When Trust Is Lost: The Impact of Interpersonal Trauma on Social Interactions," *Psychological Medicine* 49, no. 6 (2018): 1041–46.

tool of the status quo: Daniel Nettle and Rebecca Saxe, "'If Men Were Angels, No Government Would Be Necessary': The Intuitive Theory of Social Motivation and Preference for Authoritarian Leaders," *Collabra* 7, no. 1 (2021): 28105.

how divided we've become: Luiza A. Santos et al., "Belief in the Utility of Cross-Partisan Empathy Reduces Partisan Animosity and Facilitates Political Persuasion," *Psychological Science* 33, no. 9 (2022): 1557–73.

fits in perfectly with the other curses: Douglas Cairns, "Can We Find Hope in Ancient Greek Philosophy? *Elpis* in Plato and Aristotle," in *Emotions Across Cultures: Ancient China and Greece*, ed. David Konstan (Berlin: De Gruyter, 2022), 41–74.

Chapter 1: Signs and Symptoms

the Diogenes Club: Arthur Conan Doyle, *The Adventure of the Greek Interpreter*, 1893.

Diogenes of Sinope: Diogenes was not the first cynic; that title belongs to Antisthenes, a student of Socrates. However, Diogenes popularized the philosophy and is its most famous standard-bearer today. See Ansgar Allen, *Cynicism* (Cambridge, MA: MIT Press, 2020); Arthur C. Brooks, "We've Lost the True Meaning of Cynicism," *Atlantic*, May 23, 2022.

"I fawn on those who give": Diogenes, *Sayings and Anecdotes: With Other Popular Moralists*, trans. Robin Hard (Oxford: Oxford University Press, 2012).

"big-C Cynicism": Much information on Diogenes and big-C Cynicism is drawn from the writings of Luis E. Navia, especially *Diogenes the Cynic: The War Against the World* (Amherst, NY: Humanities Press, 2005); and Navia, *Classical Cynicism: A Critical Study* (New York: Bloomsbury, 1996).

Zen master slapping: A. Jesse Jiryu Davis, "Why Did Master Rinzai Slap Jo?," *EmptySquare Blog*, April 15, 2018.

what one expert calls a "missionary zeal": John Moles, "'Honestius Quam Ambitiosius'? An Exploration of the Cynic's Attitude to Moral Corruption in His Fellow Men," *Journal of Hellenic Studies* 103 (1983): 103–23.

"persuasive charm": As cited in Allen, *Cynicism*, 73–74.

their hope for humanity was left behind: For more on the evolution of Cynicism to cynicism, see Allen, *Cynicism*; David Mazella, *The Making of Modern Cynicism* (Charlottesville: University of Virginia Press, 2007); Navia, *Classical Cynicism*.

cynicism detector: For the original scale, see Walter Wheeler Cook and Donald M. Medley, "Proposed Hostility and Pharisaic-Virtue Scales for the MMPI," *Journal of Applied Psychology* 38, no. 6 (1954): 414–18. For more on its history and characteristics, see John C. Barefoot et al., "The Cook-Medley Hostility Scale: Item Content and Ability to Predict Survival," *Psychosomatic Medicine* 51, no. 1 (1989): 46–57; Timothy W. Smith and Karl D. Frohm, "What's So Unhealthy About Hostility? Construct Validity and Psychosocial Correlates of the Cook and Medley Ho Scale," *Health Psychology* 4, no. 6 (1985): 503–20.

Cook and Medley's fifty prompts: John C. Barefoot et al., "Hostility Patterns and Health Implications: Correlates of Cook-Medley Hostility Scale Scores in a National Survey," *Health Psychology* 10, no. 1 (1991): 18–24.

Optimists pay attention to good omens: Shelley E. Taylor and Jonathon D. Brown, "Illusion and Well-Being: A Social Psychological Perspective on Mental Health," *Psychological Bulletin* 103, no. 2 (1988): 193–210.

people are selfish, greedy, and dishonest: We write about this view in Eric Neumann and Jamil Zaki, "Toward a Social Psychology of Cynicism," *Trends in Cognitive Sciences* 27, no. 1 (2023): 1–3. Many other scholars have shared similar perspectives, including Kwok Leung et al., "Social Axioms," *Journal of Cross-Cultural Psychology* 33, no. 3 (2002): 286–302; Morris Rosenberg, "Misanthropy and Political Ideology," *American Sociological Review* 21, no. 6 (1956): 690–95; Lawrence S. Wrightsman, "Measurement of Philosophies of Human Nature," *Psychological Reports* 14, no. 3 (1964): 743–51.

found listeners aloof and callous: Clayton R. Critcher and David Dunning, "No Good Deed Goes Unquestioned: Cynical Reconstruals Maintain Belief in the Power of Self-Interest," *Journal of Experimental Social Psychology* 47, no. 6 (2011): 1207–13; Ana-Maria Vranceanu, Linda C. Gallo, and Laura M. Bogart, "Hostility and Perceptions of Support in Ambiguous Social Interactions," *Journal of Individual Differences* 27, no. 2 (2006): 108–15. These cynical perceptions are especially prevalent among individuals in power, who suspect their colleagues, friends, and even spouses of being out to get them; see M. Ena Inesi, Deborah H. Gruenfeld, and Adam D. Galinsky, "How Power Corrupts Relationships: Cynical Attributions for Others' Generous Acts," *Journal of Experimental Social Psychology* 48, no. 4 (2012): 795–803.

measure *trust*: A slightly more formal, classic definition from the behavioral sciences for *trust* is "a psychological state comprising the intention to accept vulnerability based upon positive expectations of the intention or behavior of another." See Denise M. Rousseau et al., "Not So Different After All: A Cross-Discipline View of Trust," *Academy of Management Review* 23, no. 3 (1998): 393–404.

everyone wins: Karen S. Cook et al., "Trust Building via Risk Taking: A Cross-Societal Experiment," *Social Psychology Quarterly* 68, no. 2 (2005): 121–42.

If you're like the average cynic: Jenny Kurman, "What I Do and What I Think They Would Do: Social Axioms and Behaviour," *European Journal of Personality* 25, no. 6 (2011): 410–23; Theodore M. Singelis et al., "Convergent Validation of the Social Axioms Survey," *Personality and Individual Differences* 34, no. 2 (2003): 269–82.

"chemically engineered": Kurt Vonnegut, *Wampeters, Foma, and Granfalloons (Opinions)* (New York: Delacorte Press, 1974).

negotiate as if the other party is trying to cheat: Singelis et al., "Convergent Validation of the Social Axioms Survey"; Pedro Neves, "Organizational Cynicism: Spillover Effects on Supervisor–Subordinate Relationships and Performance," *Leadership Quarterly* 23, no. 5 (2012): 965–76; Chia-Jung Tsay, Lisa L. Shu, and Max H. Bazerman, "Naïveté and Cynicism in Negotiations and Other Competitive Contexts," *Academy of Management Annals* 5, no. 1 (2011): 495–518.

more likely to drink: Heponiemi et al., "Longitudinal Effects of Social Support"; Siegler et al., "Patterns of Change in Hostility."

cynics financially flatline: Stavrova and Ehlebracht, "Cynical Beliefs About Human Nature and Income."

more than twice as likely: Of course, none of these studies *make* people into cynics and then measure how their lives unfold. Cynicism correlates with hardship, but we can't know from this work alone that one causes the other. But the relationship between cynicism and poor outcomes also can't be explained away by people's race, gender, or income. Cynicism alone might not ruin our lives, but it doesn't sweeten them much, either. See Everson et al., "Hostility and Increased Risk of Mortality." For more work in a similar vein, see John C. Barefoot, W. Grant Dahlstrom, and Redford B. Williams, "Hostility, CHD Incidence, and Total Mortality: A 25-Year Follow-Up Study of 255 Physicians," *Psychosomatic Medicine* 45, no. 1 (1983): 59–63; Jerry Suls, "Anger and the Heart: Perspectives on Cardiac Risk, Mechanisms and Interventions," *Progress in Cardiovascular Diseases* 55, no. 6 (2013): 538–47.

compare the well-being of high- and low-trust nations: Esteban Ortiz-Ospina, "Trust," Our World in Data, July 22, 2016.

physically healthier and more tolerant: John F. Helliwell, Haifang Huang, and Shun Wang, "New Evidence on Trust and Well-Being" (working paper, National Bureau of Economic Research, July 1, 2016).

less likely to die by suicide: John F. Helliwell, "Well-Being and Social Capital: Does Suicide Pose a Puzzle?," *Social Indicators Research* 81, no. 3 (2006): 455–96; John F. Helliwell and Shun Wang, "Trust and Well-Being" (working paper, Research Papers in Economics, April 1, 2010).

trust levels in forty-one nations: Paul J. Zak and Stephen Knack, "Trust and Growth," *Economic Journal* 111, no. 470 (2001): 295–321.

lacked...the trusting connections: Etsuko Yasui, "Community Vulnerability and Capacity in Post-Disaster Recovery: The Cases of Mano and Mikura Neighbourhoods in the Wake of the 1995 Kobe Earthquake" (PhD diss., University of British Columbia, 2007).

almost three out of every four: See table 7.3 in Yasui, "Community Vulnerability and Capacity in Post-Disaster Recovery," 226.

effect of trust isn't confined to those two neighborhoods: For instance, Kobe is split into nine areas, or "wards." Wards that were high in trust before the quake rebuilt and repopulated more quickly than low-trust ones; see Daniel P. Aldrich, "The Power of People: Social Capital's Role in Recovery from the 1995 Kobe Earthquake," *Natural Hazards* 56, no. 3 (2010): 595–611.

connections between people predict: John F. Helliwell et al., eds., *World Happiness Report 2022* (New York: Sustainable Development Solutions Network, 2022).

Crime, polarization, and disease rise: Zak and Knack, "Trust and Growth"; Jacob Dearmon and Kevin Grier, "Trust and Development," *Journal of Economic Behavior and Organization* 71, no. 2 (2009): 210–20; Oguzhan C. Dincer and Eric M. Uslaner, "Trust and Growth," *Public Choice* 142, no. 1–2 (2009): 59–67.

people's faith in government fell in the US and many other countries: In particular, surveys conducted before and during the pandemic in the US and three European nations found that people's satisfaction with their country's political system, national pride, and their support for democracy all fell during 2020. See Alexander Bor et al., "The Covid-19 Pandemic Eroded System Support but Not Social Solidarity," *PLOS ONE* 18, no. 8 (2023).

more than 80 percent of eligible South Koreans: Dasl Yoon, "Highly Vaccinated South Korea Can't Slow Down COVID-19," *Wall Street Journal*, December 16, 2021.

As Prime Minister Chung Sye-kyun later reflected: Kristen de Groot, "South Korea's Response to COVID-19: Lessons for the Next Pandemic," *Penn Today*, October 14, 2022.

more infection and death among low-trust nations: Henrikas Bartusevičius et al., "The Psychological Burden of the COVID-19 Pandemic Is Associated with Antisystemic Attitudes and Political Violence," *Psychological Science* 32, no. 9 (2021): 1391–403; Marie Fly Lindholt et al., "Public Acceptance of COVID-19 Vaccines: Cross-National Evidence on Levels and Individual-Level Predictors Using Observational Data," *BMJ Open* 11, no. 6 (2021): e048172.

40 percent of global infection could have been prevented: Thomas J. Bollyky et al., "Pandemic Preparedness and COVID-19: An Exploratory Analysis of Infection and Fatality Rates, and Contextual Factors Associated with Preparedness in 177 Countries, from Jan 1, 2020, to Sept 30, 2021," *Lancet* 399, no. 10334 (2022): 1489–1512.

"until I became a father. That changed everything": William Litster Bruneau, *The Bidet: Everything There Is to Know from the First and Only Book on the Bidet, an Elegant Solution for Comfort, Health, Happiness, Ecology, and Economy* (self-pub., 2020), 196.

Bill raised the boy alone: These biographical details drawn from Stephanie Bruneau's writing.

"The remarkable gift my father gave me": Emile Bruneau, "Atticus and Parenting—Clarity" (Word document shared by Stephanie Bruneau, March 4, 2023).

"because he was happy with nothing": Janet Lewis, a member of the rugby team whom Emile coached and his frequent travel companion, personal communication, December 13, 2022.

fed up with its glitzy fundraisers: Jeff Freund, a rugby teammate and fraternity brother of Emile's, personal communication, December 13, 2022.

spent years examining slices of brain tissue: For instance, see Emile Bruneau et al., "Increased Expression of Glutaminase and Glutamine Synthetase mRNA in the Thalamus in Schizophrenia," *Schizophrenia Research* 75, no. 1 (2005): 27–34.

brains of Palestinians and Israelis: Emile Bruneau et al., "Denying Humanity: The Distinct Neural Correlates of Blatant Dehumanization," *Journal of Experimental Psychology: General* 147, no. 7 (2018): 1078–93; Emile Bruneau, Nicholas Dufour, and Rebecca Saxe, "Social Cognition in Members of Conflict Groups: Behavioural and Neural Responses in Arabs, Israelis and South Americans to Each Other's Misfortunes," *Philosophical Transactions of the Royal Society B: Biological Sciences* 367, no. 1589 (2012): 717–30.

rarely hurried and enjoyed getting lost: Janet Lewis, personal communication, December 13, 2022.

"He wasn't a person you could 'manage'": Emily Falk, Emile's academic colleague and mentor, personal communication, December 6, 2022.

define themselves through social comparison: Alan Fontana et al., "Cynical Mistrust and the Search for Self-Worth," *Journal of Psychosomatic Research* 33, no. 4 (1989): 449–56.

more open to information that contradicts: Geoffrey L. Cohen, Joshua Aronson, and Claude M. Steele, "When Beliefs Yield to Evidence: Reducing Biased Evaluation by Affirming the Self," *Personality and Social Psychology Bulletin* 26, no. 9 (2000): 1151–64; Joshua Correll, Steven J. Spencer, and Mark P. Zanna, "An Affirmed Self and an Open Mind: Self-Affirmation and Sensitivity to Argument Strength," *Journal of Experimental Social Psychology* 40, no. 3 (2004): 350–56.

values-affirmation also increases kindness: Sander Thomaes et al., "Arousing 'Gentle Passions' in Young Adolescents: Sustained Experimental Effects of Value Affirmations on Prosocial Feelings and Behaviors," *Developmental Psychology* 48, no. 1 (2012): 103–10.

Chapter 2: The Surprising Wisdom of Hope

choose a cynic or a non-cynic: Olga Stavrova and Daniel Ehlebracht, "The Cynical Genius Illusion: Exploring and Debunking Lay Beliefs About Cynicism and Competence," *Personality and Social Psychology Bulletin* 45, no. 2 (2019): 254–69.

better at spotting liars: Carter and Weber, "Not Pollyannas." People don't just use cynicism as a clue about someone's smarts; they also use smarts as a clue about their cynicism. In a study, people guessed that competent individuals would be unfriendly, and incompetent ones would be warm and fuzzy; see Charles M. Judd et al., "Fundamental Dimensions of Social Judgment: Understanding the Relations Between Judgments of Competence and Warmth," *Journal of Personality and Social Psychology* 89, no. 6 (2005): 899–913.

gloomiest version of themselves to impress others: Deborah Son Holoien and Susan T. Fiske, "Downplaying Positive Impressions: Compensation Between Warmth and

Competence in Impression Management," *Journal of Experimental Social Psychology* 49, no. 1 (2013): 33–41.

cynics scored *less* well: See studies 4 to 6 in Stavrova and Ehlebracht, "Cynical Genius Illusion." Another project followed more than ten thousand British children over several years and found that more-intelligent kids tended to become less-cynical adults. This doesn't reflect privilege. When students arrive at college, they're just as cynical as people who don't go. Education itself opens people to trust. For more, see Toshio Yamagishi, *Trust: The Evolutionary Game of Mind and Society* (New York: Springer Science & Business Media, 2011); Noah Carl and Francesco C. Billari, "Generalized Trust and Intelligence in the United States," *PLOS ONE* 9, no. 3 (2014): e91786; Olga Stavrova and Daniel Ehlebracht, "Education as an Antidote to Cynicism," *Social Psychological and Personality Science* 9, no. 1 (2017): 59–69; Patrick Sturgis, Sanna Read, and Nick Allum, "Does Intelligence Foster Generalized Trust? An Empirical Test Using the UK Birth Cohort Studies," *Intelligence* 38, no. 1 (2010): 45–54.

terrible at picking lie *detectors*: Carter and Weber, "Not Pollyannas."

cynical children ended up the worst off: Ken J. Rotenberg, Michael J. Boulton, and Claire L. Fox, "Cross-Sectional and Longitudinal Relations Among Children's Trust Beliefs, Psychological Maladjustment and Social Relationships: Are Very High as Well as Very Low Trusting Children at Risk?," *Journal of Abnormal Child Psychology* 33, no. 5 (2005): 595–610.

Cynics work for the prosecution: An entertaining version of this very trial is held in the first episode of *Star Trek: The Next Generation*, as an all-powerful being prosecutes humanity for its many moral failures, while Patrick Stewart's Jean-Luc Picard argues for the defense.

wisdom arrives when people know what they *don't* know: Intellectual humility is only one component of wisdom; others include taking other people's perspectives and seeking out knowledge. See Mengxi Dong, Nic M. Weststrate, and Marc A. Fournier, "Thirty Years of Psychological Wisdom Research: What We Know About the Correlates of an Ancient Concept," *Perspectives on Psychological Science* 18, no. 4 (2022): 778–811; Igor Grossmann et al., "The Science of Wisdom in a Polarized World: Knowns and Unknowns," *Psychological Inquiry* 31, no. 2 (2020): 103–33.

adjust to a complex world: My use of curious skepticism is aligned with research on both intellectual humility and wisdom. See Tenelle Porter et al., "Predictors and Consequences of Intellectual Humility," *Nature Reviews Psychology* 1, no. 9 (2022): 524–36; Grossmann et al., "Science of Wisdom in a Polarized World."

more likely to fall for conspiracy theories: D. Alan Bensley et al., "Skepticism, Cynicism, and Cognitive Style Predictors of the Generality of Unsubstantiated Belief," *Applied Cognitive Psychology* 36, no. 1 (2022): 83–99. See also Büşra Elif Yelbuz, Ecesu Madan, and Sinan Alper, "Reflective Thinking Predicts Lower Conspiracy Beliefs: A Meta-Analysis," *Judgment and Decision Making* 17, no. 4 (2022): 720–44.

Megan: I've changed names and identifying details to protect Megan's identity. I interviewed Megan on March 29, 2022, and January 6, 2023, and corroborated details of her story through review of her 2020 social media posts.

compensate for those threat[s]: Karen M. Douglas et al., "Understanding Conspiracy Theories," *Political Psychology* 40, no. S1 (2019): 3–35.

conspiracy theorists tend to be more anxious: Jakub Šrol, Eva Ballová Mikušková, and Vladimíra Čavojová, "When We Are Worried, What Are We Thinking? Anxiety, Lack of Control, and Conspiracy Beliefs Amidst the COVID-19 Pandemic," *Applied Cognitive Psychology* 35, no. 3 (2021): 720–29; Ricky Green and Karen M. Douglas, "Anxious Attachment and Belief in Conspiracy Theories," *Personality and Individual Differences* 125 (2018): 30–37.

children reacted in very different ways: Mary D. Salter Ainsworth et al., *Patterns of Attachment: A Psychological Study of the Strange Situation* (Mahwah, NJ: Lawrence Erlbaum, 1978).

instability reverberates through their lives: Mario Mikulincer, "Attachment Working Models and the Sense of Trust: An Exploration of Interaction Goals and Affect Regulation," *Journal of Personality and Social Psychology* 74, no. 5 (1998): 1209–24.

more likely to distrust: Kenneth N. Levy and Benjamin N. Johnson, "Attachment and Psychotherapy: Implications from Empirical Research," *Canadian Psychology* 60, no. 3 (2019): 178–93; Anton Philipp Martinez et al., "Mistrust and Negative Self-Esteem: Two Paths from Attachment Styles to Paranoia," *Psychology and Psychotherapy* 94, no. 3 (2020): 391–406.

insecure attachment has spread: Sara Konrath et al., "Changes in Adult Attachment Styles in American College Students over Time," *Personality and Social Psychology Review* 18, no. 4 (2014): 326–48.

Insecure attachment comes in different forms: R. Chris Fraley et al., "The Experiences in Close Relationships—Relationship Structures Questionnaire: A Method for Assessing Attachment Orientations Across Relationships," *Psychological Assessment* 23, no. 3 (2011): 615–25; Chris R. Fraley and Glenn I. Roisman, "The Development of Adult Attachment Styles: Four Lessons," *Current Opinion in Psychology* 25 (2019): 26–30; Jaakko Tammilehto et al., "Dynamics of Attachment and Emotion Regulation in Daily Life: Uni- and Bidirectional Associations," *Cognition & Emotion* 36, no. 6 (2022): 1109–31.

remain skittish around new people: This example is described in an excellent paper about overgeneralization: Brian van Meurs et al., "Maladaptive Behavioral Consequences of Conditioned Fear-Generalization: A Pronounced, Yet Sparsely Studied, Feature of Anxiety Pathology," *Behaviour Research and Therapy* 57 (2014): 29–37.

destroy a person's trust: Dark moments also topple our most cherished beliefs about the world. People were once kind; now they're cruel. The world was safe; now it's perilous. One of the most popular accounts of PTSD is that it shatters and reshapes people's lay theories in this way; see Ronnie Janoff-Bulman, "Assumptive Worlds and the Stress of Traumatic Events: Applications of the Schema Construct," *Social Cognition* 7, no. 2 (1989): 113–36; Ronnie Janoff-Bulman, *Shattered Assumptions* (New York: Simon & Schuster, 2010).

to defend themselves: Mario Bogdanov et al., "Acute Psychosocial Stress Increases Cognitive-Effort Avoidance," *Psychological Science* 32, no. 9 (2021): 1463–75.

negative feedback loop: Scientists call this psychological trap a "wicked learning environment." For more, see Robin M. Hogarth, Tomás Lejarraga, and Emre Soyer, "The Two Settings of Kind and Wicked Learning Environments," *Current Directions in Psychological Science* 24, no. 5 (2015): 379–85.

Linda died when Emile was in his thirties: Stephanie Bruneau, personal communication, March 4, 2023.

"through no fault of their own": Janet Lewis, personal communication, December 13, 2022.

more likely to help strangers: Michal Bauer et al., "Can War Foster Cooperation?," *Journal of Economic Perspectives* 30, no. 3 (2016): 249–74; Patricia A. Frazier, Amy Conlon, and Theresa Glaser, "Positive and Negative Life Changes Following Sexual Assault," *Journal of Consulting and Clinical Psychology* 69, no. 6 (2001): 1048–55; Daniel Lim and David DeSteno, "Suffering and Compassion: The Links Among Adverse Life Experiences, Empathy, Compassion, and Prosocial Behavior," *Emotion* 16, no. 2 (2016): 175–82.

better chance of growing through hardship: Matthew L. Brooks et al., "Trauma Characteristics and Posttraumatic Growth: The Mediating Role of Avoidance Coping, Intrusive Thoughts, and Social Support," *Psychological Trauma: Theory, Research, Practice, and Policy* 11, no. 2 (2019): 232–38; Sarah E. Ullman and Liana C. Peter-Hagene, "Social Reactions to Sexual Assault Disclosure, Coping, Perceived Control, and PTSD Symptoms in Sexual Assault Victims," *Journal of Community Psychology* 42, no. 4 (2014): 495–508.

parents met each kid where they were: Emile Bruneau, "Atticus and Parenting—Clarity."

"earned attachment": Ximena B. Arriaga et al., "Revising Working Models Across Time: Relationship Situations That Enhance Attachment Security," *Personality and Social Psychology Review* 22, no. 1 (2017): 71–96; Atina Manvelian, "Creating a Safe Haven and Secure Base: A Feasibility and Pilot Study of Emotionally Focused Mentoring to Enhance Attachment Security" (PhD diss., University of Arizona, 2021).

build trust: For an example of trust increasing through "emotion focused therapy," see Stephanie A. Wiebe et al., "Predicting Follow-Up Outcomes in Emotionally Focused Couple Therapy: The Role of Change in Trust, Relationship-Specific Attachment, and Emotional Engagement," *Journal of Marital and Family Therapy* 43, no. 2 (2016): 213–26.

more open-minded: Matthew Jarvinen, "Attachment and Cognitive Openness: Emotional Underpinnings of Intellectual Humility," *Journal of Positive Psychology* 12, no. 1 (2016): 74–86.

openness can help others feel safe: Julia A. Minson and Frances S. Chen, "Receptiveness to Opposing Views: Conceptualization and Integrative Review," *Personality and Social Psychology Review* 26, no. 2 (2021): 93–111; Harry T. Reis et al., "Perceived Partner Responsiveness Promotes Intellectual Humility," *Journal of Experimental Social Psychology* 79 (2018): 21–33.

"motivation to look more deeply": Megan's experience here jibes with research on how motivation affects our thinking. When you expect to talk with someone who supports a political position, you read writing on that position favorably; when you expect a conversation with a critic, you pick up on critical information. Megan's "shared reality" switched, from one shared with other QAnons to one shared with Thomas. For more, see Gerald Echterhoff, E. Tory Higgins, and John M. Levine, "Shared Reality: Experiencing Commonality with Others' Inner States About the World," *Perspectives on Psychological Science* 4, no. 5 (2009): 496–521; Ziva Kunda, "The Case for Motivated Reasoning," *Psychological Bulletin* 108, no. 3 (1990): 480–98; Philip E. Tetlock and Jae Il Kim, "Accountability and Judgment Processes in a Personality Prediction Task," *Journal of Personality and Social Psychology* 52, no. 4 (1987): 700–709.

contradicts black-and-white assumptions: Yulia Landa et al., "Group Cognitive Behavioral Therapy for Delusions: Helping Patients Improve Reality Testing," *Journal of Contemporary Psychotherapy* 36, no. 1 (2006): 9–17.

Chapter 3: Preexisting Conditions

just needed a plane ticket: Andreas Leibbrandt, interviews with author, October 27, 2022, and August 30, 2023.

trust game: Uri Gneezy, Andreas Leibbrandt, and John A. List, "Ode to the Sea: Workplace Organizations and Norms of Cooperation," *Economic Journal* 126, no. 595 (2015): 1856–83.

"wilting orchid disease": Credit to psychologist Sanah Ahsan for this evocative metaphor. Sanah Ahsan, "I'm a Psychologist—and I Believe We've Been Told Devastating Lies About Mental Health," *Guardian*, September 6, 2022.

people grow kinder or crueler: This is a major theme of my last book: Jamil Zaki, *The War for Kindness: Building Empathy in a Fractured World* (New York: Crown, 2019).

Cynicism runs in families: Dorit Carmelli, Gary E. Swan, and Ray H. Rosenman, "The Heritability of the Cook and Medley Hostility Scale Revised," *Journal of Social Behavior and Personality* 5, no. 1 (1990): 107–16; Sarah S. Knox et al., "A Genome Scan for Hostility: The National Heart, Lung, and Blood Institute Family Heart Study," *Molecular Psychiatry* 9, no. 2 (2003): 124–26.

the American middle class owned about 50 percent: For interactive graphs on inequality, visit World Inequality Database, "USA," accessed October 13, 2023, https://wid.world/country/usa/. Specific calculations: In 1980, the middle 40 percent owned 30.6 percent of the wealth, and the top 1 percent owned 20.9 percent, for a ratio of 1.46:1. In 2020, the middle 40 percent owned 27.8 percent, compared to the top 1 percent's 34.9 percent. For a fuller picture of these trends and their historical context, see Thomas Piketty, *A Brief History of Equality* (Cambridge, MA: Harvard University Press, 2022).

stranded millions of people on the economic margins: For data on economic mobility, see Raj Chetty et al., "The fading American dream: Trends in absolute income mobility since 1940," *Science* 356 (2017): 398–406. For affordability of homes and education, see Stella Sechopoulos, "Most in the U.S. say young adults today face more challenges than their parents' generation in some key areas," Pew Research Center, February 28, 2022.

hostile, stressed, lonely, materialistic: Jolanda Jetten et al., "A Social Identity Analysis of Responses to Economic Inequality," *Current Opinion in Psychology* 18 (2017): 1–5; Lora

E. Park et al., "Psychological Pathways Linking Income Inequality in Adolescence to Well-Being in Adulthood," *Self and Identity* 20, no. 8 (2020): 982–1014.

mistrustful: Frank J. Elgar, "Income Inequality, Trust, and Population Health in 33 Countries," *American Journal of Public Health* 100, no. 11 (2010): 2311–15; Jolanda Jetten, Kim Peters, and Bruno Gabriel Salvador Casara, "Economic Inequality and Conspiracy Theories," *Current Opinion in Psychology* 47 (2022): 101358.

cynicism typically follows: For instance, an analysis of nineteen wealthy countries over a thirty-year span found that as nations became less equal, they also became less trusting over time. See Guglielmo Barone and Sauro Mocetti, "Inequality and Trust: New Evidence from Panel Data," *Economic Inquiry* 54, no. 2 (December 8, 2015): 794–809.

"Stasi was there and watching": Richard J. Popplewell, "The Stasi and the East German Revolution of 1989," *Contemporary European History* 1, no. 1 (1992): 37–63.

their reign left a long civic hangover: Andreas Lichter, Max Löffler, and Sebastian Siegloch, "The Long-Term Costs of Government Surveillance: Insights from Stasi Spying in East Germany," *Journal of the European Economic Association* 19, no. 2 (2020): 741–89.

less democratic between 2019 and 2020: Sarah Repucci and Amy Slipowitz, *Freedom in the World 2021: Democracy Under Siege* (New York: Freedom House, 2021).

He discovered that democracy scores have fallen: Jacob Grumbach, "Laboratories of Democratic Backsliding," *American Political Science Review* 117, no. 3 (2023): 967–84.

Before the "Partygate" scandal: Andrew Woodcock, "Trust in Politicians 'in Freefall' After Year of Chaos at Westminster," *Independent*, November 3, 2022.

In the era of WeWork: As we've seen, cynics generally perform less well than non-cynics on cognitive tests. But in highly corrupt nations, clever people are much more likely to be cynical. When elites abuse their power, cynicism becomes more common—and more reasonable. See Stavrova and Ehlebracht, "Cynical Genius Illusion."

twenty-two thousand emergencies across the city were not called in: Matthew Desmond, Andrew V. Papachristos, and David S. Kirk, "Police Violence and Citizen Crime Reporting in the Black Community," *American Sociological Review* 81, no. 5 (2016): 857–76; Bill McCarthy, John Hagan, and Daniel Herda, "Neighborhood Climates of Legal Cynicism and Complaints About Abuse of Police Power," *Criminology* 58, no. 3 (2020): 510–36.

Pre-disappointment can also worsen health disparities: Itai Bavli and David S. Jones, "Race Correction and the X-Ray Machine—The Controversy over Increased Radiation Doses for Black Americans in 1968," *New England Journal of Medicine* 387, no. 10 (2022): 947–52; Kelly M. Hoffman et al., "Racial Bias in Pain Assessment and Treatment Recommendations, and False Beliefs About Biological Differences Between Blacks and Whites," *Proceedings of the National Academy of Sciences* 113, no. 16 (2016): 4296–301.

Many worried that vaccine appointments would turn into immigration ambushes: Chris Iglesias of Fruitvale's Unity Council, interview with author, January 27, 2023.

only about 65 percent of Fruitvale: This is a conservative estimate of the discrepancy. In fact, by May 2021, 77 percent of the entire area code, including Piedmont and less-wealthy areas, had been vaccinated. By August of that year, a full third of Fruitvale residents remained unvaccinated. See Brian Krans, "How Flaws in California's Vaccine System Left Some Oaklanders Behind," *Oaklandside*, May 18, 2021. Deepa Fernandes, "Children of Immigrants at the Heart of Effort to Reach Oakland's Unvaccinated Communities," *San Francisco Chronicle*, August 11, 2021.

eight times more likely: Erin E. Esaryk et al., "Variation in SARS-COV-2 Infection Risk and Socioeconomic Disadvantage Among a Mayan-Latinx Population in Oakland, California," *JAMA Network Open* 4, no. 5 (2021): e2110789.

Exchange relationships: Data on this come from around the world. See, for instance, Heidi Colleran, "Market Integration Reduces Kin Density in Women's Ego-Networks in Rural Poland," *Nature Communications* 11, no. 1 (2020): 1–9; Robert Thomson et al., "Relational Mobility Predicts Social Behaviors in 39 Countries and Is Tied to Historical Farming and Threat," *Proceedings of the National Academy of Sciences* 115, no. 29 (2018): 7521–26; Kristopher M Smith, Ibrahim A. Mabulla, and Coren L. Apicella, "Hadza Hunter-Gatherers with Greater Exposure to Other Cultures Share More with Generous Campmates," *Biology Letters* 18, no. 7 (2022): 20220157. For a general theory of social marketplaces and cooperation, see Pat Barclay, "Biological Markets and the Effects of Partner Choice on Cooperation and Friendship," *Current Opinion in Psychology* 7 (2016): 33–38.

Markets are driven by self-interest: Paul Lodder et al., "A Comprehensive Meta-Analysis of Money Priming," *Journal of Experimental Psychology: General* 148, no. 4 (2019): 688–712.

Exchange is fine for business: Ryan W. Carlson and Jamil Zaki, "Good Deeds Gone Bad: Lay Theories of Altruism and Selfishness," *Journal of Experimental Social Psychology* 75 (2018): 36–40.

A date whom you pay for their time: As the philosopher Michael Sandel puts it, "The money that would buy the friendship dissolves the good I seek to acquire." Michael Sandel, "Market Reasoning as Moral Reasoning: Why Economists Should Re-Engage with Political Philosophy," *Journal of Economic Perspectives* 27, no. 4 (2013): 121–40.

"increase the risk of addiction": Jaclyn Smock, "Smartwatches Can Be Toxic, Too" *Allure*, October 14, 2022.

not because they *feel* tired: Kelly Glazer Baron et al., "Orthosomnia: Are Some Patients Taking the Quantified Self Too Far?," *Journal of Clinical Sleep Medicine* 13, no. 2 (2017): 351–54.

"the social media in-person collision": Quotes drawn from Lane's interview with First Person; see Lulu Garcia-Navarro et al., "The Teenager Leading the Smartphone Liberation Movement," opinion, *New York Times*, February 2, 2023.

visited counseling services more often: Luca Braghieri, Ro'ee Levy, and Alexey Makarin, "Social Media and Mental Health," *American Economic Review* 112, no. 11 (2022): 3660–93.

"identify the people in your life who score highest": Tara Parker-Pope, "The Power of Positive People," *New York Times*, July 12, 2018.

dwarfing any other form of matchmaking: Michael Rosenfeld, Reuben J. Thomas, and Sonia Hausen, "Disintermediating Your Friends: How Online Dating in the United States Displaces Other Ways of Meeting," *Proceedings of the National Academy of Sciences* 116, no. 36 (2019): 17753–58.

Tinder founders modeled their app on slot machines: See interview with Tinder cofounder Jonathan Badeen in the *Verge*'s podcast about dating apps: Sangeeta Singh-Kurtz, "How Tinder Changed Everything," *Verge*, January 11, 2023.

weighing suitors' statistics against one another: Gabriel Bonilla-Zorita, Mark D. Griffiths, and Daria J. Kuss, "Online Dating and Problematic Use: A Systematic Review," *International Journal of Mental Health and Addiction* 19, no. 6 (2020): 2245–78.

According to psychologist Steven Pinker: Steven Pinker, *Enlightenment Now* (New York: Viking Press, 2018).

unequal, corrupt, and commodified settings all raise anomie: Ali Teymoori, Brock Bastian, and Jolanda Jetten, "Towards a Psychological Analysis of Anomie," *Political Psychology* 38, no. 6 (2016): 1009–23; Lea Hartwich and Julia Becker, "Exposure to Neoliberalism Increases Resentment of the Elite via Feelings of Anomie and Negative Psychological Reactions," *Journal of Social Issues* 75, no. 1 (2019): 113–33; Karim Bettache, Chi-yue Chiu, and Peter Beattie, "The Merciless Mind in a Dog-Eat-Dog Society: Neoliberalism and the Indifference to Social Inequality," *Current Opinion in Behavioral Sciences* 34 (2020): 217–22; Jetten, Peters, and Casara, "Economic Inequality and Conspiracy Theories."

On weekends, father and son worked as janitors: Source on Bill and Emile's story: Stephanie Bruneau. During Emile's time at the school, most of Peninsula's janitorial staff was made up of parents. The barter policy changed in the 1990s and now all staff are paid. Source: Andromeda Garcelon, Emile's Peninsula classmate and now a parent and community ambassador for the school. Interviewed on January 18, 2023.

"high on cooperation and creativity": As per a document titled "Monica Meyer," sent by Emile to Janet Lewis on July 16, 2019.

trust children: Scientists don't know if trusting parents *caused* kids to put faith in others, but any caregiver can test this out themselves by showing kids we believe in them. See Dan Wang and Anne C. Fletcher, "Parenting Style and Peer Trust in Relation to School Adjustment in Middle Childhood," *Journal of Child and Family Studies* 25, no. 3 (2015): 988–98.

pursued for their own sake: This is consistent with the philosopher Kieran Setiya's suggestion to pursue "atelic" activies, free of an end goal or purpose; see Kieran Setiya, *Midlife: A Philosophical Guide* (Princeton, NJ: Princeton University Press, 2017).

become happier and less stressed: Morelli et al., "Emotional and Instrumental Support Provision."

digital cleanse: Hunt Allcott et al., "The Welfare Effects of Social Media," *American Economic Review* 110, no. 3 (2020): 629–76.

view the world as dangerous: Jeremy D. W. Clifton and Peter Meindl, "Parents Think—Incorrectly—That Teaching Their Children That the World Is a Bad Place Is Likely Best for Them," *Journal of Positive Psychology* 17, no. 2 (2021): 182–97.

"be cautious": Dietlind Stolle and Laura Nishikawa, "Trusting Others—How Parents Shape the Generalized Trust of Their Children," *Comparative Sociology* 10, no. 2 (2011): 281–314.

making Generation Z the least trusting: Jean M. Twenge, W. Keith Campbell, and Nathan T. Carter, "Declines in Trust in Others and Confidence in Institutions Among American Adults and Late Adolescents, 1972–2012," *Psychological Science* 25, no. 10 (2014): 1914–23.

savoring: For a great, comprehensive view of this practice and its benefits, see Fred B. Bryant and Joseph Veroff, *Savoring: A New Model of Positive Experience* (New York: Psychology Press, 2017).

Filipinos trust: See table 7 in Martin Paldam, "Social Capital and Social Policy" (working paper, New Frontiers of Social Policy: Development in a Globalizing World, 2005), 11.

vaccinate fifteen thousand people in 2021: Data obtained from conversation with the Unity Council's chief operating officer, Armando Hernandez, interview with author, May 1, 2023. For more on the campaign, see Brian Krans, "'We Are in a Race': With Delta Variant Cases Spiking, Oakland Continues Vaccination Push," *Oaklandside*, August 6, 2021; Leonardo Castañeda, "'In the Trenches': Students Walk the Streets of Hard-Hit Fruitvale Seeking COVID Vaccine Holdouts," *Daily Democrat*, July 3, 2021.

conservatives watch Republican politicians: Katherine Clayton and Robb Willer, "Endorsements from Republican Politicians Can Increase Confidence in U.S. Elections," *Research & Politics* 10, no. 1 (2023): 205316802211489; Sophia Pink et al., "Elite Party Cues Increase Vaccination Intentions Among Republicans," *Proceedings of the National Academy of Sciences* 118, no. 32 (2021): e2106559118.

Chapter 4: Hell Isn't Other People

wallets were returned: For the original "experiment," see Diana Zlomislic, "We Left 20 Wallets Around the GTA. Most Came Back," *Toronto Star*, April 25, 2009; and Helliwell and Wang, "Trust and Well-Being." A larger study of more than seventeen thousand dropped wallets in forty countries found that Canada is toward the high end of global trustworthiness, but the majority of wallets around the world were returned. Interestingly, they were *more* likely to be returned if they had money in them, versus no money, again undercutting the stereotype that people are selfish. See Alain Cohn et al., "Civic Honesty Around the Globe," *Science* 365, no. 6448 (2019): 70–73.

cheater detection: For a short review, see Leda Cosmides et al., "Detecting Cheaters," *Trends in Cognitive Sciences* 9, no. 11 (2005): 505–6.

volunteering, donating to charity, and helping strangers all *increased* significantly during the pandemic: John F. Helliwell et al., eds., *World Happiness Report 2023*, 11th ed. (New York: Sustainable Development Solutions Network, 2023).

barely a quarter noticed the vast COVID kindness: Jamil Zaki, "The COVID Kindness We Ignored" (unpublished essay).

people regularly fail to realize: Cameron Brick et al., "Self-Interest Is Overestimated: Two Successful Pre-Registered Replications and Extensions of Miller and Ratner (1998),"

Collabra Psychology 7, no. 1 (2021): 23443; Dale T. Miller and Rebecca K. Ratner, "The Disparity Between the Actual and Assumed Power of Self-Interest," *Journal of Personality and Social Psychology* 74, no. 1 (1998): 53–62; Nicholas Epley and David Dunning, "Feeling 'Holier Than Thou': Are Self-Serving Assessments Produced by Errors in Self- or Social Prediction?," *Journal of Personality and Social Psychology* 79, no. 6 (2000): 861–75; Nicholas Epley et al., "Undersociality: Miscalibrated Social Cognition Can Inhibit Social Connection," *Trends in Cognitive Sciences* 26, no. 5 (2022): 406–18; Dale T. Miller, "The Norm of Self-Interest," *American Psychologist* 54, no. 12 (1999): 1053–60; Detlef Fetchenhauer and David Dunning, "Do People Trust Too Much or Too Little?," *Journal of Economic Psychology* 30, no. 3 (2009): 263–76.

"Troubles kick our door in and come and find us": Fred Bryant, "You 2.0: Slow Down!," interview by Shankar Vedantam (Hidden Brain Media, n.d.).

Negativity bias: John J. Skowronski and Donal E. Carlston, "Negativity and Extremity Biases in Impression Formation: A Review of Explanations," *Psychological Bulletin* 105, no. 1 (1989): 131–42.

humanity is in a state of vicious decline: Adam Mastroianni and Daniel T. Gilbert, "The Illusion of Moral Decline," *Nature* 618, no. 7966 (2023): 782–89.

cooperated 9 percent *more* across time: Mingliang Yuan et al., "Did Cooperation Among Strangers Decline in the United States? A Cross-Temporal Meta-Analysis of Social Dilemmas (1956–2017)," *Psychological Bulletin* 148, no. 3–4 (2022): 129–57.

***gossip*:** Robin Dunbar, Anna Marriott, and Neill Duncan, "Human Conversational Behavior," *Human Nature* 8, no. 3 (1997): 231–46.

Fearing shame and retribution: Matthew Feinberg, Robb Willer, and Michael Schultz, "Gossip and Ostracism Promote Cooperation in Groups," *Psychological Science* 25, no. 3 (2014): 656–64; Manfred Milinski, Dirk Semmann, and Hans-Jürgen Krambeck, "Reputation Helps Solve the 'Tragedy of the Commons,'" *Nature* 415, no. 6870 (2002): 424–26.

People who read these notes, in turn: Samantha Grayson et al., "Gossip Decreases Cheating but Increases (Inaccurate) Cynicism" (manuscript in preparation).

"society will get better": David Bornstein and Tina Rosenberg, "When Reportage Turns to Cynicism," opinion, *New York Times*, November 14, 2016.

***each* negative word in a headline increased its number of views:** Claire Robertson et al., "Negativity Drives Online News Consumption," *Nature Human Behaviour* 7, no. 5 (2023): 812–22.

growing presence of negative emotions: David Rozado, Ruth Hughes, and Jamin Halberstadt, "Longitudinal Analysis of Sentiment and Emotion in News Media Headlines Using Automated Labelling with Transformer Language Models," *PLOS ONE* 17, no. 10 (2022): e0276367.

"hate" tripled: Charlotte Olivia Brand, Alberto Acerbi, and Alex Mesoudi, "Cultural Evolution of Emotional Expression in 50 Years of Song Lyrics," *Evolutionary Human Sciences* 1 (January 1, 2019): E1.

algorithms surround us with more of what we fear and loathe: If consumers shape the media, media shapes our worldview right back. In one classic experiment, researchers paid people to tamper with their news diet. The scientists chose an issue, such as rising taxes or carbon emissions, and snuck a two-minute segment on that topic into viewers' programming each night for one week. You probably think your sense of what matters is shaped over a lifetime of experience. But if you're like the people in this study, it might take just twelve minutes. Adding that much coverage on an issue made it vastly more likely that viewers later said it was *the most important* one facing the nation. See Shanto Iyengar and Donald R. Kinder, *News That Matters: Television and American Opinion*, updated ed. (Chicago: University of Chicago Press, 2010).

most people thought crime had increased: Data drawn from Justin McCarthy, "Perceptions of Increased U.S. Crime at Highest Since 1993," Gallup, November 20, 2021. The 25/27 figure indicates the years in which more than 50 percent of respondents answered that there was "more" crime in the US than a year ago.

actual crime rate decreased by nearly 50 percent: "Reported Violent Crime Rate in the U.S. 2021," Statista, retrieved October 10, 2023.

"crime wave": Annie Lowrey, "Why San Francisco Prosecutor Chesa Boudin Faces Recall," *Atlantic*, May 20, 2022.

most likely to believe crime is ascendant: Valerie J. Callanan, "Media Consumption, Perceptions of Crime Risk and Fear of Crime: Examining Race/Ethnic Differences," *Sociological Perspectives* 55, no. 1 (2012): 93–115.

"you'd be bracing for it all the time": Quote drawn from the podcast *How To*. Transcript available at Nicole Lewis and Amanda Ripley, "How to Unbreak the News," *Slate*, August 30, 2022.

An ice sheet the size of Delaware broke off Antarctica: Sean Greene, "Antarctica Shed a Block of Ice the Size of Delaware, but Scientists Think the Real Disaster Could Be Decades Away," *Los Angeles Times*, January 20, 2018.

US has closed nearly seventeen hundred polling places: "Democracy Diverted: Polling Place Closures and the Right to Vote," Leadership Conference on Civil and Human Rights, September 10, 2019, https://civilrights.org/democracy-diverted/.

actively avoid media: Nic Newman, "Overview and Key Findings of the 2022 Digital News Report," Reuters Institute for the Study of Journalism, June 15, 2022.

stop shoveling bad news: Solutions Journalism, "The Top 10 Takeaways from the Newest Solutions Journalism Research," *Medium—The Whole Story*, January 6, 2022.

"off planet": Quotes here drawn from Shorters's interview in the podcast *On Being*; see Trabian Shorters, "Trabian Shorters—A Cognitive Skill to Magnify Humanity," interview by Krista Tippett (The On Being Project, 2022).

"true self": George Newman, Paul Bloom, and Joshua Knobe, "Value Judgments and the True Self," *Personality and Social Psychology Bulletin* 40, no. 2 (2013): 203–16.

"good true self" effect: Julian De Freitas et al., "Origins of the Belief in Good True Selves," *Trends in Cognitive Sciences* 21, no. 9 (2017): 634–36; De Freitas et al., "Consistent Belief in a Good True Self in Misanthropes and Three Interdependent Cultures," *Cognitive Science* 42, no. 51 (2013): 134–60.

crime in *their area* had stayed the same or lessened: McCarthy, "Perceptions of Increased U.S. Crime at Highest Since 1993."

70 percent of them opted to read negative stories: Toni G. L. A. Van Der Meer and Michael Hameleers, "I Knew It, the World Is Falling Apart! Combatting a Confirmatory Negativity Bias in Audiences' News Selection Through News Media Literacy Interventions," *Digital Journalism* 10, no. 3 (2022): 473–92.

comparable to small business loans in the US: Board of Governors of the Federal Reserve System (US), "Delinquency Rate on Business Loans, All Commercial Banks," FRED, Federal Reserve Bank of St. Louis, accessed October 15, 2023.

"The only images I had of Bangladeshi villagers": Quotes drawn from Lewis and Ripley, "How to Unbreak the News."

simplistic and helpless: Amy J. C. Cuddy, Mindi S. Rock, and Michael I. Norton, "Aid in the Aftermath of Hurricane Katrina: Inferences of Secondary Emotions and Intergroup Helping," *Group Processes & Intergroup Relations* 10, no. 1 (2007): 107–18.

"positive deviants": David Bornstein and Tina Rosenberg, "11 Years of Lessons from Reporting on Solutions," opinion, *New York Times*, November 11, 2021.

"Often I'm depressed for half the day": David Byrne, "Reasons to Be Cheerful," David Byrne, 2018, https://davidbyrne.com/explore/reasons-to-be-cheerful/about.

Women Overcoming Recidivism Through Hard Work: Maurice Chammah, "To Help Young Women in Prison, Try Dignity," opinion, *New York Times*, October 9, 2018.

TeleHelp Ukraine: MaryLou Costa, "The World's Therapists Are Talking to Ukraine," *Reasons to Be Cheerful*, August 25, 2023.

the Solutions Journalism Network (SJN): "Solutions Story Tracker®," Solutions Journalism, accessed October 15, 2023, https://www.solutionsjournalism.org/storytracker.

"green bank": Ashley Stimpson, "'Green Banks' Are Turning Climate Action Dreams into Realities," *Reasons to Be Cheerful*, December 21, 2022.

Florida ballot initiative: Jenna Spinelle, "For the Many or the Few?," Solutions Journalism, August 1, 2022.

balance negative conversation with celebration: Research finds that in economic games, "positive gossip" is just as useful in promoting kind behavior as the negative kind; see Hirotaka Imada, Tim Hopthrow, and Dominic Abrams, "The Role of Positive and Negative Gossip in Promoting Prosocial Behavior," *Evolutionary Behavioral Sciences* 15, no. 3 (2021): 285–91.

Chapter 5: Escaping the Cynicism Trap

the *Boston Globe* published a scathing exposé: David Armstrong, "Money to Burn," *Boston Globe*, Sunday, February 7, 1999.

an "alarming" number of injuries: Sarah Schweitzer, "City, Firefighters Settle," *Boston Globe*, August 31, 2001.

"tough enough to work through fatigue or illness": "The Boston Globe 05 Jul 2002, Page 5," Boston Globe Archive, accessed October 15, 2023.

"When you're injured, you can't function": Schweitzer, "City, Firefighters Settle."

"several ugly scenes": Douglas Belkin, "Uncertainty for Fire Dept. Reform," *Boston Globe*, September 27, 2001.

The chief promised to investigate: Scott Greenberger, "Fire Head Suspends 18 Over Sick Pay," *Boston Globe*, July 11, 2003.

backfired spectacularly: Greenberger, "Fire Head Suspends 18."

Accused of being selfish: For more on this story from a behavioral science perspective, see Samuel Bowles, *The Moral Economy: Why Good Incentives Are No Substitute for Good Citizens* (New Haven, CT: Yale University Press, 2016); and Tess Wilkinson-Ryan, "Do Liquidated Damages Encourage Breach? A Psychological Experiment," *Michigan Law Review* 108 (2010): 633–72.

more likely to eavesdrop: Jennifer Carson Marr et al., "Do I Want to Know? How the Motivation to Acquire Relationship-Threatening Information in Groups Contributes to Paranoid Thought, Suspicion Behavior, and Social Rejection," *Organizational Behavior and Human Decision Processes* 117, no. 2 (2012): 285–97.

tend toward emotional abuse: Geraldine Downey and Scott I. Feldman, "Implications of Rejection Sensitivity for Intimate Relationships," *Journal of Personality and Social Psychology* 70, no. 6 (1996): 1327–43; Lindsey M. Rodriguez et al., "The Price of Distrust: Trust, Anxious Attachment, Jealousy, and Partner Abuse," *Partner Abuse* 6, no. 3 (2015): 298–319.

disappear when others need them: Seth A. Kaplan, Jill C. Bradley, and Janet B. Ruscher, "The Inhibitory Role of Cynical Disposition in the Provision and Receipt of Social Support: The Case of the September 11th Terrorist Attacks," *Personality and Individual Differences* 37, no. 6 (2004): 1221–32.

"influence neglect": For a terrific summary of this work, see Vanessa K. Bohns, *You Have More Influence Than You Think: How We Underestimate Our Powers of Persuasion, and Why It Matters* (Washington, DC: National Geographic Books, 2023).

more than half did: Vanessa K. Bohns, M. Mahdi Roghanizad, and Amy Z. Xu, "Underestimating Our Influence over Others' Unethical Behavior and Decisions," *Personality and Social Psychology Bulletin* 40, no. 3 (2013): 348–62; Francis J. Flynn and Vanessa K. B. Lake, "If You Need Help, Just Ask: Underestimating Compliance with Direct Requests for Help," *Journal of Personality and Social Psychology* 95, no. 1 (2008): 128–43.

people asked strangers to vandalize a library book: Bohns, Roghanizad, and Xu, "Underestimating Our Influence over Others' Unethical Behavior and Decisions."

when investors sent more money, trustees sent back more: Noel D. Johnson and Alexandra Mislin, "Trust Games: A Meta-Analysis," *Journal of Economic Psychology* 32, no. 5 (2011): 865–89. Figures are calculated as follows: Average investment = 50 percent. Average repayment = 37 percent. Standard deviation of investment = .12, meaning that an investment of 62 percent is one standard deviation above the mean. According to Johnson and Mislin, "A one standard deviation increase in trust, leads to about a 40 percent increase in trustworthiness." A 40 percent increase over the mean repayment rate would come to ~52 percent returned.

"I don't believe in you": The same is true when investors try to force trustees to repay, for instance, by putting contingencies and penalties alongside their investments. See Armin Falk and Michael Kosfeld, "The Hidden Costs of Control," *American Economic Review* 96, no. 5 (2006): 1611–30.

"**I *do* believe in you**": Ernesto Reuben, Paola Sapienza, and Luigi Zingales, "Is Mistrust Self-Fulfilling?," *Economics Letters* 104, no. 2 (2009): 89–91.

self-fulfilling prophecies: Marr et al., "Do I Want to Know?"

lose interest in the relationship: Downey and Feldman, "Implications of Rejection Sensitivity for Intimate Relationships."

making it more likely their friends will actually disrespect them: Olga Stavrova, Daniel Ehlebracht, and Kathleen D. Vohs, "Victims, Perpetrators, or Both? The Vicious Cycle of Disrespect and Cynical Beliefs About Human Nature," *Journal of Experimental Psychology: General* 149, no. 9 (2020): 1736–54.

cynics decide they were right all along: In nerdier terms, preemptive strikes are one example of a "wicked learning environment," in which the evidence people learn from is biased, thus driving them toward systematically wrong conclusions; see Robin M. Hogarth, Tomás Lejarraga, and Emre Soyer, "Two Settings of Kind and Wicked Learning Environments."

"**Firefighters have shown that their putative pride doesn't make them immune to that impulse**": Columnist Scott Lehigh, quoted from "The Boston Globe 16 Jul 2003, Page 19," Boston Globe Archive, accessed October 15, 2023.

Generous Tit for Tat: Robert M. Axelrod and Douglas Dion, "The Further Evolution of Cooperation," *Science* 242, no. 4884 (1988): 1385–90; Jian Wu and Robert Axelrod, "How to Cope with Noise in the Iterated Prisoner's Dilemma," *Journal of Conflict Resolution* 39, no. 1 (1995): 183–89.

"**the property of *being nice***": Robert M. Axelrod, *The Evolution of Cooperation* (New York: Basic Books, 1984), 33.

"**the future must cast a sufficiently large shadow**": Robert Axelrod, "The Evolution of Cooperation," Stanford University Department of Electrical Engineering, 1984, accessed October 15, 2023, https://ee.stanford.edu/~hellman/Breakthrough/book/chapters/axelrod.html#Live.

Recently, my lab tested whether teaching people about their power: Eric Neumann et al., "People Trust More After Learning Trust Is Self-Fulfilling" (manuscript in preparation).

"**showed up one day buff**": Andromeda Garcelon, interview with author, January 31, 2023.

injury rate three times higher: Nienke W. Willigenburg et al., "Comparison of Injuries in American Collegiate Football and Club Rugby," *American Journal of Sports Medicine* 44, no. 3 (2016): 753–60.

"**cloud nine from the sheer intensity**": Franck Boivert, interview with author, February 17, 2023.

"**I'm with you**": Quotes drawn from a document Emile emailed to Janet Lewis on July 16, 2019, titled "Monica Mayer." He wrote to Janet, "Here's that piece that I'm not sure I'll be able to fit into the book . . . but a story that I want to tell somehow."

"**If you have fun, you're always alert**": Boivert continues to beat the drum of unconventional practice, now living and coaching in Fiji. As he recently said, "Coaches should be developing the intelligence of players instead of making them play like robots." See Meli Laddpeter, "Being 'Franck': Players Need to Be Allowed to Play Freely—Boivert," *Fiji Times*, May 11, 2022.

"That left room for us to relax and play": Janet Lewis, interview with author, December 13, 2022; Janet Lewis, email correspondence, May 30, 2023.

precisely that rash, uncounting quality that makes trust most powerful: Jutta Weber, Deepak Malhotra, and J. Keith Murnighan, "Normal Acts of Irrational Trust: Motivated Attributions and the Trust Development Process," *Research in Organizational Behavior* 26 (2004): 75–101.

invest in others in "uncalculating" ways: Jillian J. Jordan et al., "Uncalculating Cooperation Is Used to Signal Trustworthiness," *Proceedings of the National Academy of Sciences* 113, no. 31 (2016): 8658–63.

unilateral move: One important caveat to this story is that the Americans, in part through intelligence gathered by Soviet double agents, knew that the USSR was overstating its nuclear capabilities. As such, JFK's speech might be perceived as calling Khrushchev's bluff rather than offering peace. Nonetheless, doing so through a unilateral offer of de-escalation allowed the Soviets to save face while increasing the likelihood of peace.

It was a game of generous tit for tat on a global stage: Svenn Lindskold, "Trust Development, the GRIT Proposal, and the Effects of Conciliatory Acts on Conflict and Cooperation," *Psychological Bulletin* 85, no. 4 (1978): 772–93.

Chapter 6: The (Social) Water Is Just Fine

hikikomori: Watanabe's story was drawn from email interviews conducted between October and December 2022, and from a photo essay he published online about his experience, which can be found at https://dajf.org.uk/wp-content/uploads/Atsushi-Watanabe-presentation.pdf.

rankings of artists' quality and popularity: Art Compass and Artifacts provide two such rankings for individual artists.

one in every hundred Japanese adults: Takahiro A. Kato, Shigenobu Kanba, and Alan R. Teo, "Hikikomori: Multidimensional Understanding, Assessment, and Future International Perspectives," *Psychiatry and Clinical Neurosciences* 73, no. 8 (2019): 427–40.

reported in Spain, Oman, and the US: Tanner J. Bommersbach and Hun Millard, "No Longer Culture-Bound: Hikikomori Outside of Japan," *International Journal of Social Psychiatry* 65, no. 6 (2019): 539–40.

nearly 1 percent of adults in many nations live in near-total isolation: Alan Teo, PhD, interview with author, March 28, 2023.

fivefold increase in just two decades: Daniel A. Cox, "Men's Social Circles Are Shrinking," Survey Center on American Life, June 29, 2021.

reported feeling lonely: Jean M. Twenge et al., "Worldwide Increases in Adolescent Loneliness," *Journal of Adolescence* 93 (2021): 257–69.

intensifies depression: The best single treatment of loneliness remains the late John Cacioppo's modern classic; see John T. Cacioppo and William H. Patrick, *Loneliness: Human Nature and the Need for Social Connection* (New York: W. W. Norton, 2008).

caught colds more often: Sheldon Cohen, "Social Relationships and Health," *American Psychologist* 59, no. 8 (2004): 676–84; Sheldon Cohen et al., "Social Ties and Susceptibility to the Common Cold," *JAMA* 277, no. 24 (1997): 1940–44.

increased mortality risk: Julianne Holt-Lunstad et al., "Loneliness and Social Isolation as Risk Factors for Mortality," *Perspectives on Psychological Science* 10, no. 2 (2015): 227–37.

"ever-increasing price": Office of the Surgeon General, *Our Epidemic of Loneliness and Isolation: The U.S. Surgeon General's Advisory on the Healing Effects of Social Connection and Community* (Washington, DC: U.S. Public Health Service, 2023), 4.

compared social predictions to reality: Nicholas Epley and Juliana Schroeder, "Mistakenly Seeking Solitude," *Journal of Experimental Psychology: General* 143, no. 5 (2014): 1980–99; Juliana Schroeder, Donald W. Lyons, and Nicholas Epley, "Hello, Stranger? Pleasant Conversations Are Preceded by Concerns About Starting One," *Journal of Experimental Psychology: General* 151, no. 5 (2022): 1141–53.

usually glad to help: Xuan Zhao and Nicholas Epley, "Surprisingly Happy to Have Helped: Underestimating Prosociality Creates a Misplaced Barrier to Asking for Help," *Psychological Science* 33, no. 10 (2022): 1708–31.

boost people's mood and draw us closer: Erica J. Boothby and Vanessa K. Bohns, "Why a Simple Act of Kindness Is Not as Simple as It Seems: Underestimating the Positive Impact of Our Compliments on Others," *Personality and Social Psychology Bulletin* 47, no. 5 (2020): 826–40; Amit Kumar and Nicholas Epley, "Undervaluing Gratitude: Expressers Misunderstand the Consequences of Showing Appreciation," *Psychological Science* 29, no. 9 (2018): 1423–35.

extroversion reported being happier: The positive benefits of acting extroverted are largest for extroverts, while introverts also reported more fatigue when doing so. But both groups still saw mood benefits. For more work on this, see William Fleeson, Adriane B. Malanos, and Noelle M. Achille, "An Intraindividual Process Approach to the Relationship Between Extraversion and Positive Affect: Is Acting Extraverted as 'Good' as Being Extraverted?," *Journal of Personality and Social Psychology* 83, no. 6 (2002): 1409–22; Rowan Jacques-Hamilton, Jessie Sun, and Luke D. Smillie, "Costs and Benefits of Acting Extraverted: A Randomized Controlled Trial," *Journal of Experimental Psychology: General* 148, no. 9 (2019): 1538–56; Seth Margolis and Sonja Lyubomirsky, "Experimental Manipulation of Extraverted and Introverted Behavior and Its Effects on Well-Being," *Journal of Experimental Psychology: General* 149, no. 4 (2020): 719–31; John M. Zelenski, Maya S. Santoro, and Deanna C. Whelan, "Would Introverts Be Better Off If They Acted More Like Extraverts? Exploring Emotional and Cognitive Consequences of Counterdispositional Behavior," *Emotion* 12, no. 2 (2012): 290–303.

"perception that people are wronging me": Alan Teo, interview with author, March 28, 2023.

others don't want or need us: Among the *hikikomori*, anxiety drives people inside, and aloneness makes them feel less prepared to reemerge. Saito Tamaki, a pioneer in the study of people suffering with this condition, writes about its vicious cycle: "In ordinary diseases, when an individual grows sick, their bodies will react naturally with various therapeutic measures, including immune responses...In the case of withdrawal, however, the unhealthy state has the function of making the situation even worse." See Saitō Tamaki, *Hikikomori: Adolescence Without End*, trans. Jeffrey Angles (Minneapolis: University of Minnesota Press, 2013), 81.

with physical complaints: Tegan Cruwys et al., "Social Isolation Predicts Frequent Attendance in Primary Care," *Annals of Behavioral Medicine* 52, no. 10 (February 3, 2018): 817–29; Fuschia M. Sirois and Janine Owens, "A Meta-Analysis of Loneliness and Use of Primary Health Care," *Health Psychology Review* 17, no. 2 (2021): 193–210.

symptoms spiked: Akram Parandeh et al., "Prevalence of Burnout Among Health Care Workers During Coronavirus Disease (COVID-19) Pandemic: A Systematic Review and Meta-Analysis," *Professional Psychology: Research and Practice* 53, no. 6 (2022): 564–73; H. J. A. Van Bakel et al., "Parental Burnout Across the Globe During the COVID-19 Pandemic," *International Perspectives in Psychology* 11, no. 3 (2022): 141–52.

self-care industry: Jamil Zaki, "We Should Try Caring for Others as 'Self-Care,'" *Atlantic*, October 21, 2021.

burned-out people grow more cynical: Christina Maslach and Michael P. Leiter, "Understanding the Burnout Experience: Recent Research and Its Implications for Psychiatry," *World Psychiatry* 15, no. 2 (2016): 103–11; Christina Maslach, Wilmar B. Schaufeli, and Michael P. Leiter, "Job Burnout," *Annual Review of Psychology* 52, no. 1 (2001): 397–422.

protects against distress and exhaustion: Shauna L. Shapiro, Kirk Warren Brown, and Gina M. Biegel, "Teaching Self-Care to Caregivers: Effects of Mindfulness-Based Stress Reduction on the Mental Health of Therapists in Training," *Training and Education in Professional Psychology* 1, no. 2 (2007): 105–15.

they often feel replenished: Frank Martela and Richard M. Ryan, "The Benefits of Benevolence: Basic Psychological Needs, Beneficence, and the Enhancement of Well-Being," *Journal of Personality* 84, no. 6 (2015): 750–64; Jason D. Runyan et al., "Using Experience Sampling to Examine Links Between Compassion, Eudaimonia, and Pro-Social Behavior," *Journal of Personality* 87, no. 3 (2018): 690–701.

feel less depressed: Bruce Doré et al., "Helping Others Regulate Emotion Predicts Increased Regulation of One's Own Emotions and Decreased Symptoms of Depression," *Personality and Social Psychology Bulletin* 43, no. 5 (2017): 729–39; Morelli et al., "Emotional and Instrumental Support Provision."

compassion toward others: Kira Schabram and Yu Tse Heng, "How Other- and Self-Compassion Reduce Burnout Through Resource Replenishment," *Academy of Management Journal* 65, no. 2 (2022): 453–78.

people tend to ignore the good news: Elizabeth W. Dunn, Lara B. Aknin, and Michael I. Norton, "Spending Money on Others Promotes Happiness," *Science* 319, no. 5870 (2008): 1687–88; Cassie Mogilner, Zoë Chance, and Michael I. Norton, "Giving Time Gives You Time," *Psychological Science* 23, no. 10 (2012): 1233–38.

the worse their depression became: Jennifer Crocker et al., "Interpersonal Goals and Change in Anxiety and Dysphoria in First-Semester College Students," *Journal of Personality and Social Psychology* 98, no. 6 (2010): 1009–24.

Naikan replaces negative assumptions: For more, see Gregg Krech, *Naikan: Gratitude, Grace, and the Japanese Art of Self-Reflection* (Berkeley, CA: Stone Bridge Press, 2022).

"conversation scavenger hunt": Gillian M. Sandstrom, Erica J. Boothby, and Gus Cooney, "Talking to Strangers: A Week-Long Intervention Reduces Psychological Barriers to Social Connection," *Journal of Experimental Social Psychology* 102 (2022): 104356. You can find the scavenger hunt instructions at Gillian M. Sandstrom, "Scavenger Hunt Missions," Gillian Sandstrom, April 2021, https://gilliansandstrom.files.wordpress.com/2021/04/scavenger-hunt-missions.pdf.

increased sense of connection, meaning, and well-being: Julia Vera Pescheny, Gurch Randhawa, and Yannis Pappas, "The Impact of Social Prescribing Services on Service Users: A Systematic Review of the Evidence," *European Journal of Public Health* 30, no. 4 (2019): 664–73.

social prescribing decreased patients' loneliness: Adam Jeyes and Laura Pugh, "Implementation of Social Prescribing to Reduce Frequent Attender Consultation Rates in Primary Care," *British Journal of General Practice* 69, no. S1 (2019). For great coverage of social prescribing, see Julia Hotz, "A Radical Plan to Treat Covid's Mental Health Fallout," *WIRED UK*, August 18, 2021.

animals had to tend to one another: P. A. Kropotkin, "Mutual Aid a Factor of Evolution," *Political Science Quarterly* 18, no. 4 (1903): 702–5. For writing about his life, see James Hamlin, "Who Was…Peter Kropotkin?," *Biologist*, accessed October 15, 2023, https://www.rsb.org.uk/biologist-features/who-was-peter-kropothkin; Lee Alan Dugatkin, "The Prince of Evolution: Peter Kropotkin's Adventures in Science and Politics," *Scientific American*, September 13, 2011.

"survival programs": Black Panther Party Legacy & Alumni, "Survival Programs," It's About Time, accessed October 15, 2023.

"Caring for myself is not self-indulgence": Audre Lorde, *A Burst of Light: And Other Essays* (Mineola, NY: Courier Dover, 2017).

self-care industrial complex loses the term's original meaning: Lenora E. Houseworth, "The Radical History of Self-Care," *Teen Vogue*, January 14, 2021; Aimaloghi Eromosele, "There Is No Self-Care Without Community Care," *URGE—Unite for Reproductive & Gender Equity* (blog), November 10, 2020; Aisha Harris, "How 'Self-Care' Went from Radical to Frou-Frou to Radical Once Again," *Slate*, April 5, 2017.

Mutual aid programs: Jia Tolentino, "What Mutual Aid Can Do During a Pandemic," *New Yorker*, May 11, 2020; Sigal Samuel, "Coronavirus Volunteering: How You Can Help Through a Mutual Aid Group," *Vox*, April 16, 2020.

"Coffee Break Project": Cassady Rosenblum and September Dawn Bottoms, "How Farmers in Colorado Are Taking Care of Their Mental Health," *New York Times*, October 15, 2022.

Chapter 7: Building Cultures of Trust

"Why You Don't Want to Be Microsoft CEO": Dina Bass, "Microsoft CEO: World's Worst Job," *Bloomberg News*, January 30, 2014.

Microsoft had lost half its value: This and several other details drawn from Kurt Eichenwald's excellent *Vanity Fair* profile of Microsoft's "lost decade": Kurt Eichenwald,

"How Microsoft Lost Its Mojo: Steve Ballmer and Corporate America's Most Spectacular Decline," *Vanity Fair*, July 24, 2012.

Organizational cynicism: James W. Dean, Pamela Brandes, and Ravi Dharwadkar, "Organizational Cynicism," *Academy of Management Review* 23, no. 2 (1998): 341–52.

"an employer's most productive asset": This quote and most coverage of GE and Jack Welch here is drawn from the excellent book by David Gelles: *The Man Who Broke Capitalism: How Jack Welch Gutted the Heartland and Crushed the Soul of Corporate America—and How to Undo His Legacy* (New York: Simon & Schuster, 2022).

***homo economicus*:** Joseph Persky, "Retrospectives: The Ethology of *Homo Economicus*," *Journal of Economic Perspectives* 9, no. 2 (1995): 221–31.

more principled than *economicus*: For a classic in this debunking, see Amartya Sen, "Rational Fools: A Critique of the Behavioral Foundations of Economic Theory," *Philosophy & Public Affairs* 6, no. 4 (1977): 317–44.

believe in fundamental selfishness: Amitaï Etzioni, "The Moral Effects of Economic Teaching," *Sociological Forum* 30, no. 1 (2015): 228–33; Robert H. Frank, Thomas D. Gilovich, and Dennis T. Regan, "Do Economists Make Bad Citizens?," *Journal of Economic Perspectives* 10, no. 1 (1996): 187–92.

"The growth of a large business is merely a survival of the fittest": John J. Dwyer, "Darwinism and Populism," *John J Dwyer* (blog), April 1, 2022.

"companies must compete not only with their competitors": Sumantra Ghoshal, "Bad Management Theories Are Destroying Good Management Practices," *Academy of Management Learning and Education* 4, no. 1 (2005): 75–91.

Microsoft also went to war: Eichenwald, "How Microsoft Lost Its Mojo."

acquiring Nokia in 2013: Matt Rosoff, "Satya Nadella Just Undid Steve Ballmer's Last Big Mistake," *Business Insider*, July 8, 2015.

a "culture of genius" that erodes trust: Elizabeth A. Canning et al., "Cultures of Genius at Work: Organizational Mindsets Predict Cultural Norms, Trust, and Commitment," *Personality and Social Psychology Bulletin* 46, no. 4 (2019): 626–42.

they use preemptive strikes: Bradley J. Alge, Gary A. Ballinger, and Stephen G. Green, "Remote Control: Predictors of Electronic Monitoring Intensity and Secrecy," *Personnel Psychology* 57, no. 2 (2004): 377–410.

"productivity points": Jodi Kantor et al., "Workplace Productivity: Are You Being Tracked?," *New York Times*, September 6, 2023.

re-chained herself to the desk: Danielle Abril and Drew Harwell, "Keystroke Tracking, Screenshots, and Facial Recognition: The Boss May Be Watching Long After the Pandemic Ends," *Washington Post*, September 27, 2021.

"ensure they didn't get ahead of me": Eichenwald, "How Microsoft Lost Its Mojo." Knowledge hoarding is rampant in cynical organizations. Susan Fowler describes very similar practices in her viral 2017 blog post about working at Uber: "One of the directors boasted to our team that he had withheld business-critical information from one of the executives so that he could curry favor with one of the other executives (and, he told us with a smile on his face, it worked!)." See Susan Fowler, "Reflecting on One Very, Very

Strange Year at Uber—Susan Fowler," *Susan Fowler Blog*, May 22, 2017, https://www.susanjfowler.com/blog/2017/2/19/reflecting-on-one-very-strange-year-at-uber.

comfort zones: Andrew Armatas, "How the Solution Becomes the Problem: The Performance Solution That Backfired at Microsoft," in *SAGE Business Cases* (Thousand Oaks, CA: SAGE Publications, 2023).

cynical jobs tend to be shorter: Rebecca Abraham, "Organizational Cynicism: Bases and Consequences," *Genetic, Social, and General Psychology Monographs* 126, no. 3 (2000): 269–92; Dan S. Chiaburu et al., "Antecedents and Consequences of Employee Organizational Cynicism: A Meta-Analysis," *Journal of Vocational Behavior* 83, no. 2 (2013): 181–97; Catherine E. Connelly et al., "Knowledge Hiding in Organizations," *Journal of Organizational Behavior* 33, no. 1 (2011): 64–88.

war intensifies generosity within groups: Bauer et al., "Can War Foster Cooperation?"; Ayelet Gneezy and Daniel M. T. Fessler, "Conflict, Sticks and Carrots: War Increases Prosocial Punishments and Rewards," *Proceedings of the Royal Society B: Biological Sciences* 279, no. 1727 (2011): 219–23. Some theorists even believe that conflict between groups encouraged the evolution of cooperation within them, through group selection pressures. For more, see Samuel Bowles, "Did Warfare Among Ancestral Hunter-Gatherers Affect the Evolution of Human Social Behaviors?," *Science* 324, no. 5932 (2009): 1293–98.

risk their lives for one another: One explanation for this extreme cooperation is "identity fusion," under which people feel as though they and their group are one; see Harvey Whitehouse et al., "The Evolution of Extreme Cooperation via Shared Dysphoric Experiences," *Scientific Reports* 7, no. 1 (2017): 1–10.

"transaction costs": David A. Lesmond, Joseph P. Ogden, and Charles Trzcinka, "A New Estimate of Transaction Costs," *Review of Financial Studies* 12, no. 5 (1999): 1113–41; Howard A. Shelanski and Peter G. Klein, "Empirical Research in Transaction Cost Economics: A Review and Assessment," *Journal of Law, Economics & Organization* 11, no. 2 (1995): 335–61.

"the lowest point of my life": McCombs School of Business, "Wells Fargo Fraud," Ethics Unwrapped, February 16, 2023, https://ethicsunwrapped.utexas.edu/video/wells-fargo-fraud.

"persistently dangerous" schools: For the New York State Education Department's criteria for persistent danger, see "Criteria for Designating Persistently Dangerous School Using SV," New York State Education Department, May 11, 2023, https://www.p12.nysed.gov/sss/ssae/schoolsafety/vadir/CriteriaforDesignatingPersistentlyDangerousSchoolusingSV.html.

"Violent or Disruptive Incident Reporting": VADIR criteria can be found through the New York State Education Department; see "SSEC—School Safety and Educational Climate," New York State Education Department, June 16, 2023, https://www.p12.nysed.gov/sss/ssae/schoolsafety/vadir/.

being "disrespectful": Jason A. Okonofua, Gregory M. Walton, and Jennifer L. Eberhardt, "A Vicious Cycle: A Social–Psychological Account of Extreme Racial Disparities in School Discipline," *Perspectives on Psychological Science* 11, no. 3 (2016): 381–98.

"engage in more defiant behaviors": Juan Del Toro et al., "The Spillover Effects of Classmates' Police Intrusion on Adolescents' School-Based Defiant Behaviors: The Mediating Role of Institutional Trust," *American Psychologist* (2023): advance online publication.

"our own people just accepted it": This and other quotes from Nadella drawn from Satya Nadella, *Hit Refresh: The Quest to Rediscover Microsoft's Soul and Imagine a Better Future for Everyone* (New York: HarperCollins, 2017).

evaluated not just on individual performance: Shana Lebowitz, "Microsoft's HR Chief Reveals How CEO Satya Nadella Is Pushing to Make Company Culture a Priority, the Mindset She Looks for in Job Candidates, and Why Individual Success Doesn't Matter as Much as It Used To," *Business Insider*, August 16, 2019.

task interdependence increases trust: Bart A. De Jong, Kurt T. Dirks, and Nicole Gillespie, "Trust and Team Performance: A Meta-Analysis of Main Effects, Moderators, and Covariates," *Journal of Applied Psychology* 101, no. 8 (2016): 1134–50; Sandy D. Staples and Jane Webster, "Exploring the Effects of Trust, Task Interdependence and Virtualness on Knowledge Sharing in Teams," *Information Systems Journal* 18, no. 6 (2008): 617–40.

expanded mental health benefits: Tom Warren, "Microsoft Employees Are Getting Unlimited Time Off," *Verge*, January 11, 2023.

2021 HR Executive of the Year: Kathryn Mayer, "How the HR Executive of the Year Rebooted Microsoft's Culture," *HR Executive*, October 6, 2021.

"This completely changes your outlook": Quotes drawn from conversations with White and her appearance on the *Om Travelers* podcast; LaJuan White, interview with author, March 17, 2022; LaJuan White, "Episode 12—LaJuan White," interview by Tyler Cagwin, January 7, 2019.

"classroom hierarchy": For details on the hierarchy at Lincoln and its connections to restorative justice, see Julie McMahon, "How a Syracuse Middle School Got Taken off State's 'Persistently Dangerous' List," *Syracuse*, August 16, 2016.

"I have to resolve [disciplinary issues] on my own": Quote drawn from Casey Quinlan, "One School District Is Fighting Decades of 'Punishment Culture,'" *Think Progress Archive*, January 30, 2017.

less likely to fall through the cracks: Jamie Amemiya, Adam Fine, and Ming Te Wang, "Trust and Discipline: Adolescents' Institutional and Teacher Trust Predict Classroom Behavioral Engagement Following Teacher Discipline," *Child Development* 91, no. 2 (2019): 661–78; Jason A. Okonofua et al., "A Scalable Empathic-Mindset Intervention Reduces Group Disparities in School Suspensions," *Science Advances* 8, no. 12 (2022): eabj0691; Jason A. Okonofua, Amanda D. Perez, and Sean Darling-Hammond, "When Policy and Psychology Meet: Mitigating the Consequences of Bias in Schools," *Science Advances* 6, no. 42 (2020): eaba9479.

employee ratings climbed about twice as fast: Jamil Zaki, Hitendra Wadhwa, and Ferose V. R., "It's Time to Teach Empathy and Trust with the Same Rigor as We Teach Coding," *Fast Company*, November 11, 2022, https://www.fastcompany.com/90808273/its-time-to-teach-empathy-and-trust-with-the-same-rigor-as-we-teach-coding.

on strike: Molly Cook Escobar and Christine Zhang, "A Summer of Strikes," *New York Times*, September 15, 2023, https://www.nytimes.com/interactive/2023/09/03/business/economy/strikes-union-sag-uaw.html.

more than two-thirds of Americans support unions: Lydia Saad, "More in U.S. See Unions Strengthening and Want It That Way," *Gallup News*, August 30, 2023, https://news.gallup.com/poll/510281/unions-strengthening.aspx.

Chapter 8: The Fault in Our Fault Lines

Soviet tanks fired into civilian buildings: Paul Lendvai, *One Day That Shook the Communist World: The 1956 Hungarian Uprising and Its Legacy* (Princeton, NJ: Princeton University Press, 2010).

Project RYaN: Bernd Schaefer, Nate Jones, and Benjamin B. Fischer, "Forecasting Nuclear War," Wilson Center, accessed October 16, 2023, https://www.wilsoncenter.org/publication/forecasting-nuclear-war.

"We've descended into civil war": Mike Giglio, "Inside the Pro-Trump Militant Group the Oath Keepers," *Atlantic*, November 2020.

69 percent of both Democrats and Republicans believed: Peter Baker and Blake Hounshell, "Parties' Divergent Realities Challenge Biden's Defense of Democracy," *New York Times*, September 2, 2022.

disliked the other side more than they liked their own: Eli J. Finkel et al., "Political Sectarianism in America," *Science* 370, no. 6516 (2020): 533–36.

Americans have "sorted": Ethan Kaplan, Jörg L. Spenkuch, and Rebecca Sullivan, "Partisan Spatial Sorting in the United States: A Theoretical and Empirical Overview," *Journal of Public Economics* 211 (2022): 104668. Sorting occurs down to the neighborhood level, with a large proportion of voters having "virtually no exposure to voters from the other party in their residential environment." See Jacob R. Brown and Ryan D. Enos, "The Measurement of Partisan Sorting for 180 Million Voters," *Nature Human Behaviour* 5, no. 8 (2021): 998–1008.

guess wrong about one another's lives: Douglas J. Ahler and Gaurav Sood, "The Parties in Our Heads: Misperceptions About Party Composition and Their Consequences," *Journal of Politics* 80, no. 3 (2018): 964–81.

Republican dog-lovers think Democrats must like cats: Kathryn R. Denning and Sara D. Hodges, "When Polarization Triggers Out-Group 'Counter-Projection' Across the Political Divide," *Personality and Social Psychology Bulletin* 48, no. 4 (2021): 638–56.

"false polarization": Matthew Levendusky and Neil Malhotra, "(Mis)Perceptions of Partisan Polarization in the American Public," *Public Opinion Quarterly* 80, no. S1 (2015): 378–91.

The more specific the questions, the more wrong we become: Stephen Hawkins et al., *Defusing the History Wars: Finding Common Ground in Teaching America's National Story* (New York: More in Common, 2022).

two landscapes: Images from "America's Divided Mind: Understanding the Psychology That Drives Us Apart," Beyond Conflict—Putting Experience and Science to Work for Peace, May 2020, https://beyondconflictint.org/americas-divided-mind/; see also

Jens Hainmueller and Daniel J. Hopkins, "The Hidden American Immigration Consensus: A Conjoint Analysis of Attitudes Toward Immigrants," *American Journal of Political Science* 59, no. 3 (2014): 529–48.

150 issues on which Republicans and Democrats agreed: University of Maryland School of Public Policy, "Major Report Shows Nearly 150 Issues on Which Majorities of Republicans & Democrats Agree," Program for Public Consultation, August 7, 2020, https://publicconsultation.org/defense-budget/major-report-shows-nearly-150-issues-on-which-majorities-of-republicans-democrats-agree/; Steve Corbin, "Americans Largely Agree on Several Key Issues and Congress Should Pay Attention," *NC Newsline*, August 19, 2022.

dreaming up rivals four times more bloodthirsty: Joseph S Mernyk et al., "Correcting Inaccurate Metaperceptions Reduces Americans' Support for Partisan Violence," *Proceedings of the National Academy of Sciences* 119, no. 16 (2022): e2116851119.

hitting one another twice as hard: Sukhwinder S. Shergill et al., "Two Eyes for an Eye: The Neuroscience of Force Escalation," *Science* 301, no. 5630 (2003): 187.

"The average Trump supporter believes": Appearance on *The Gist*, July 25, 2022; see Malcolm Nance, "Civil War: Possible or Probable?," interview by Mike Pesca.

"conflict entrepreneurs": Amanda Ripley, *High Conflict: Why We Get Trapped and How We Get Out* (New York: Simon & Schuster, 2022).

depicting political rivals: Joseph N. Cappella and Kathleen Hall Jamieson, *Spiral of Cynicism: The Press and the Public Good* (New York: Oxford University Press, 1997); Claes H. De Vreese, "The Spiral of Cynicism Reconsidered," *European Journal of Communication* 20, no. 3 (2005): 283–301.

people who felt relatively calm pretended to be outraged: William J. Brady et al., "How Social Learning Amplifies Moral Outrage Expression in Online Social Networks," *Science Advances* 7, no. 33 (2021): eabe5641; William J. Brady et al., "Overperception of Moral Outrage in Online Social Networks Inflates Beliefs About Intergroup Hostility," *Nature Human Behaviour* 7, no. 6 (2023): 917–27; William J. Brady et al., "Emotion Shapes the Diffusion of Moralized Content in Social Networks," *Proceedings of the National Academy of Sciences* 114, no. 28 (2017): 7313–18.

polarization is a major problem: Santos et al., "Belief in the Utility of Cross-Partisan Empathy."

"failed democracy": "Harvard Youth Poll," Institute of Politics at Harvard University, fall 2021, https://iop.harvard.edu/youth-poll/42nd-edition-fall-2021.

the less hope they expressed: Béatrice S. Hasler et al., "Young Generations' Hopelessness Perpetuates Long-Term Conflicts," *Scientific Reports* 13, no. 1 (2023): 1–13.

few miles from poverty and brutality: Andrés Casas, interviews with author, January 13, 2023, and September 21, 2023.

Kidnappings, rape, and torture were rampant: Nicholas Casey, "Colombia Signs Peace Agreement with FARC After 5 Decades of War," *New York Times*, September 26, 2016.

dabbling in several other fields: Andrés Casas, "Education," LinkedIn, accessed October 16, 2023, https://www.linkedin.com/in/andrescasascasas/details/education/.

Medellín, had been the world's most dangerous city: *Colombia Journal*, "Colombia: The Occupied Territories of Medellín," Relief Web, October 31, 2002, https://reliefweb

.int/report/colombia/colombia-occupied-territories-medell%C3%ADn; Ivan Erre Jota, "The Medellin Miracle," *Rapid Transition Alliance*, December 19, 2018.

"War is war, it never ends": Joe Parkin Daniels, "Colombia's Ex-Guerrillas: Isolated, Abandoned and Living in Fear," *Guardian*, February 3, 2021.

"I don't know how right I'm going to be": Quotes drawn from the Discovery Channel documentary *Why We Hate*. The episode dedicated to Emile is titled "Hope." See Emile Bruneau, "Hope" (Discovery Channel, March 11, 2019).

"I hope the police don't come knocking at my door": Samantha Moore-Berg, interview with author, January 31, 2023.

yearning for peace remained: Emile Bruneau et al., "Exposure to a Media Intervention Helps Promote Support for Peace in Colombia," *Nature Human Behaviour* 6, no. 6 (2022): 847–57.

displacing three times as many people in 2021: Norwegian Refugee Council, "Colombia: Conflict Persists Five Years After Peace Deal," NRC, November 24, 2022.

"You give me hope": Quote drawn from PirataFilms, *All It Takes*, short film; English version, *Vimeo*, August 19, 2022, https://vimeo.com/741321924.

"His wife probably wanted to kill me": Knowing Stephanie, I am almost certain Andrés is wrong here.

"We Colombians are on the brink of peace": Quote taken from PirataFilms, *All It Takes*.

same strategy has worked in subsequent studies: For instance, the Strengthening Democracy Challenge (in which I was lucky to play a tiny role) was a mega-collaboration in which social scientists brought together twenty-five interventions to reduce partisan animosity. Many of the most effective entailed simply correcting people's misperceptions about political rivals; in other words, giving them better data. See Jan G. Voelkel et al., "Megastudy Identifying Effective Interventions to Strengthen Americans' Democratic Attitudes," *OSF Preprints*, March 20, 2023.

prefer a painful dental procedure: Jeremy A. Frimer, Linda J. Skitka, and Matt Motyl, "Liberals and Conservatives Are Similarly Motivated to Avoid Exposure to One Another's Opinions," *Journal of Experimental Social Psychology* 72 (2017): 1–12.

my lab invited over a hundred Americans: Luiza Santos et al., "The Unexpected Benefits of Cross-Party Conversations on Polarized Issues" (manuscript in preparation).

conversations between rivals: Emile Bruneau et al., "Intergroup Contact Reduces Dehumanization and Meta-Dehumanization: Cross-Sectional, Longitudinal, and Quasi-Experimental Evidence from 16 Samples in Five Countries," *Personality and Social Psychology Bulletin* 47, no. 6 (2020): 906–20.

recipe for disagreeing better: See, for instance, Emily Kubin et al., "Personal Experiences Bridge Moral and Political Divides Better Than Facts," *Proceedings of the National Academy of Sciences* 118, no. 6 (2021): e2008389118; Minson and Chen, "Receptiveness to Opposing Views"; Marshall B. Rosenberg, *Nonviolent Communication: A Language of Life* (Encinitas, CA: PuddleDancer Press, 2003).

became more open-minded: Michael Yeomans et al., "Conversational Receptiveness: Improving Engagement with Opposing Views," *Organizational Behavior and Human Decision Processes* 160 (2020): 131–48.

less likely to challenge injustice, bias, and abuse: Brett Q. Ford and Allison S. Troy, "Reappraisal Reconsidered: A Closer Look at the Costs of an Acclaimed Emotion-Regulation Strategy," *Current Directions in Psychological Science* 28, no. 2 (2019): 195–203.

reducing animosity: David E. Broockman and Joshua Kalla, "Durably Reducing Transphobia: A Field Experiment on Door-to-Door Canvassing," *Science* 352, no. 6282 (2016): 220–24; Joshua Kalla and David E. Broockman, "Reducing Exclusionary Attitudes Through Interpersonal Conversation: Evidence from Three Field Experiments," *American Political Science Review* 114, no. 2 (2020): 410–25.

Chapter 9: Building the World We Want

Almost half of America's wealth was owned by the top 1 percent: World Inequality Database, "USA."

"hire one-half of the working class to shoot the other half to death": David Huyssen, "We Won't Get Out of the Second Gilded Age the Way We Got Out of the First," *Vox*, April 1, 2019.

"Some people were better at the contest of life than others": Robert D. Putnam, *The Upswing: How America Came Together a Century Ago and How We Can Do It Again* (New York: Simon & Schuster, 2020), 167.

post–Civil War racial progress reversed: Robert D. Putnam, "Bowling Alone: America's Declining Social Capital," *Journal of Democracy* 6, no. 1 (January 1, 1995): 65–78.

entice readers: David Nasaw, *The Chief: The Life of William Randolph Hearst* (New York: Houghton Mifflin Harcourt, 2000), 77.

"no control over the course of their affairs": Woodrow Wilson, "The New Freedom: A Call for the Emancipation of the Generous Energies of a People," *Political Science Quarterly* 29, no. 3 (1914): 506–7.

In 1960, 58 percent of Americans had trusted: Putnam, *Upswing*, 159.

"collect food stamps, veterans benefits ... as well as welfare": Josh Levin, "The Real Story of Linda Taylor, America's Original Welfare Queen," *Slate*, December 19, 2013.

Linda Taylor: Josh Levin provides a thorough, compelling, and tragic biography of Taylor in Josh Levin, *The Queen: The Forgotten Life Behind an American Myth* (New York: Back Bay Books, 2020).

fraud investigations jumped by more than 700 percent: Julilly Kohler-Hausmann, "'The Crime of Survival': Fraud Prosecutions, Community Surveillance, and the Original 'Welfare Queen,'" *Journal of Social History* 41, no. 2 (2007): 329–54.

Supplemental Nutrition Assistance Program (SNAP): Independent Lens, "From Mothers' Pensions to Welfare Queens, Debunking Myths About Welfare," PBS, May 16, 2023, https://www.pbs.org/independentlens/blog/from-mothers-pensions-to-welfare-queens-debunking-myths-about-welfare/. For the full report, see Daniel R. Cline and Randy Alison Aussenberg, "Errors and Fraud in the Supplemental Nutrition Assistance Program (SNAP)," Federation of American Scientists (Congressional Research Service, September 28, 2018), https://sgp.fas.org/crs/misc/R45147.pdf.

inspired public support for a 1982 bill: Levin, *Queen.*

Temporary Assistance for Needy Families program: Zachary Parolin, "Decomposing the Decline of Cash Assistance in the United States, 1993 to 2016," *Demography* 58, no. 3 (2021): 1119–41.

households living in extreme poverty: David Brady and Zachary Parolin, "The Levels and Trends in Deep and Extreme Poverty in the United States, 1993–2016," *Demography* 57, no. 6 (2020): 2337–60; Luke H. Shaefer and Kathryn Edin, "Rising Extreme Poverty in the United States and the Response of Federal Means-Tested Transfer Programs," *Social Service Review* 87, no. 2 (2013): 250–68.

the less they support welfare: Jesper Akesson et al., "Race and Redistribution in the US: An Experimental Analysis," CEPR, January 31, 2023, https://cepr.org/voxeu/columns/race-and-redistribution-us-experimental-analysis.

use child tax credits to buy drugs: Rebecca Shabad et al., "Manchin Privately Raised Concerns That Parents Would Use Child Tax Credit Checks on Drugs," *NBC News*, December 20, 2021; David Firestone, "How to Use the Debt Ceiling to Inflict Cruelty on the Poor," *New York Times*, May 17, 2023.

a "constitution for knaves": For much more on this idea, see Bowles, *Moral Economy.*

"We didn't know we were poor": William Goodwin, interviews with author, March 21, 2022, and May 3, 2023.

American children experienced poverty: "Income, Poverty and Health Insurance Coverage in the United States: 2022," United States Census Bureau, September 12, 2023, https://www.census.gov/newsroom/press-releases/2023/income-poverty-health-insurance-coverage.html. https://confrontingpoverty.org/poverty-facts-and-myths/americas-poor-are-worse-off-than-elsewhere/.

costs the US $500 billion per year: Harry J. Holzer et al., "The Economic Costs of Childhood Poverty in the United States," *Journal of Children and Poverty* 14, no. 1 (2008): 41–61.

poorest US citizens die an average of five years earlier: John Burn-Murdoch, "Why Are Americans Dying So Young?," *Financial Times*, March 31, 2023.

support public aid: Gianmarco Daniele and Benny Geys, "Interpersonal Trust and Welfare State Support," *European Journal of Political Economy* 39 (2015): 1–12. These scientists also argue for a *causal* effect of trust on social support. Specifically, children of immigrants' support for social welfare reflects the national trust of their *parents'* country of origin.

"Povertyism": OHCHR, "Ban 'Povertyism' in the Same Way as Racism and Sexism: UN Expert," October 28, 2022, https://www.ohchr.org/en/press-releases/2022/10/ban-povertyism-same-way-racism-and-sexism-un-expert.

twenty-six-page application: Annie Lowrey, "$100 Million to Cut the Time Tax," *Atlantic*, April 26, 2022.

decreases human mental capacity: Anandi Mani et al., "Poverty Impedes Cognitive Function," *Science* 341, no. 6149 (2013): 976–80.

poor Americans pass up over $13 billion in food stamps: Matthew Desmond, *Poverty, by America* (New York: Crown, 2023), 87–88.

when inequality rises, faith in people and institutions falls: Aina Gallego, "Inequality and the Erosion of Trust Among the Poor: Experimental Evidence," *Socio-Economic Review* 14, no. 3 (2016): 443–60.

tax breaks dwarfed all spending on housing assistance: Desmond, *Poverty, by America*, 91.

"not used a government social program": Henry Farrell, "The Invisible American Welfare State," *Good Authority*, February 8, 2011, https://goodauthority.org/news/the-invisible-american-welfare-state/.

Millions of people drink water that fails to meet: Emily Holden et al., "More Than 25 Million Americans Drink from the Worst Water Systems," *Consumer Reports*, February 26, 2021.

lack health insurance: Andrew P. Wilper et al., "Health Insurance and Mortality in US Adults," *American Journal of Public Health* 99, no. 12 (2009): 2289–95.

suicide: Diana E. Naranjo, Joseph E. Glass, and Emily C. Williams, "Persons with Debt Burden Are More Likely to Report Suicide Attempt Than Those Without," *Journal of Clinical Psychiatry* 82, no. 3 (2021): 31989.

"superyachts": Ralph Dazert, "Market Insight: US Continues to Dominate Superyacht Market," *SuperYacht Times*, December 1, 2022.

"No way am I worthy of recognition, let alone investment": Jesús Gerena, interview with author, February 18, 2022.

the average UpTogether family reports: Stand Together, "How Shifting Perceptions of Low-Income Families Helps Them Get Out of Poverty," https://standtogether.org/news/shifting-perceptions-of-low-income-families-is-key-to-getting-them-out-of-poverty/.

"temptation goods": David K. Evans and Anna M. Popova, "Cash Transfers and Temptation Goods: A Review of Global Evidence" (working paper, World Bank Policy Research, 2014).

spent no more on temptation goods: Data retrieved from impact report, available at "Our Impact," Foundations for Social Change, 2021, https://forsocialchange.org/impact.

saved the shelter system $8,277: Ryan Dwyer et al., "Unconditional Cash Transfers Reduce Homelessness," *Proceedings of the National Academy of Sciences* 120, no. 36 (2023): e2222103120.

earn more from farming: Katia Covarrubias, Benjamin Davis, and Paul Winters, "From Protection to Production: Productive Impacts of the Malawi Social Cash Transfer Scheme," *Journal of Development Effectiveness* 4, no. 1 (2012): 50–77; Paul Gertler, Sebastián Martínez, and Marta Rubio-Codina, "Investing Cash Transfers to Raise Long-Term Living Standards," *American Economic Journal: Applied Economics* 4, no. 1 (2012): 164–92; Johannes Haushofer and Jeremy P. Shapiro, "The Short-Term Impact of Unconditional Cash Transfers to the Poor: Experimental Evidence from Kenya," *Quarterly Journal of Economics* 131, no. 4 (2016): 1973–2042.

no major reductions in labor: Solomon Asfaw et al., "The Impact of the Kenya CT-OVC Programme on Productive Activities and Labour Allocation," *From Protection to Production Project* (Food and Agriculture Organization of the United Nations, 2013); Mouhcine Guettabi, "What Do We Know About the Effects of the Alaska Permanent Fund Dividend?," *ScholarWorks@UA* (University of Alaska Anchorage, Institute of

Social and Economic Research, 2019); Olli Kangas et al., "The Basic Income Experiment 2017–2018 in Finland: Preliminary Results," *Valto* (Ministry of Social Affairs and Health, 2019).

more prone to mental illness: David G. Weissman et al., "State-Level Macro-Economic Factors Moderate the Association of Low Income with Brain Structure and Mental Health in U.S. Children," *Nature Communications* 14, no. 1 (2023): 2085.

infants' brains developed more rapidly: Sonya V. Troller-Renfree et al., "The Impact of a Poverty Reduction Intervention on Infant Brain Activity," *Proceedings of the National Academy of Sciences* 119, no. 5 (2022): e2115649119.

Most predicted it would go toward drugs: Dwyer et al., "Unconditional Cash Transfers Reduce Homelessness."

top tax rates: "Historical Highest Marginal Income Tax Rates: 1913 to 2023," Tax Policy Center, May 11, 2023, https://www.taxpolicycenter.org/statistics/historical-highest-marginal-income-tax-rates.

Chapter 10: The Optimism of Activism

"some things in our world, to which we should never be adjusted": Martin Luther King Jr., "King's Challenge to the Nation's Social Scientists," APA, 1967, https://www.apa.org/topics/equity-diversity-inclusion/martin-luther-king-jr-challenge.

"both victims of the system and its instruments": Václav Havel, *The Power of the Powerless* (New York: Random House, 2018).

people who trust others are more likely than cynics to vote: Maria Theresia Bäck and Henrik Serup Christensen, "When Trust Matters—a Multilevel Analysis of the Effect of Generalized Trust on Political Participation in 25 European Democracies," *Journal of Civil Society* 12, no. 2 (2016): 178–97.

sign petitions, join lawful demonstrations: Michelle Benson and Thomas R. Rochon, "Interpersonal Trust and the Magnitude of Protest," *Comparative Political Studies* 37, no. 4 (2004): 435–57.

"firehose of falsehoods": Christopher Paul and Miriam Matthews, *The Russian "Firehose of Falsehood" Propaganda Model: Why It Might Work and Options to Counter It* (Santa Monica, CA: RAND Corporation, 2016).

In 2021, researchers interviewed Russians: Paul Shields, "Killing Politics Softly: Unconvincing Propaganda and Political Cynicism in Russia," *Communist and Post-Communist Studies* 54, no. 4 (2021): 54–73.

"aim of totalitarian education": Hannah Arendt, *The Origins of Totalitarianism* (New York: Houghton Mifflin Harcourt, 1973).

"Hope is a dimension of the spirit": Václav Havel, *Letters to Olga: June 1979–September 1982* (New York: Alfred A. Knopf, 1988).

he gave a sermon: This sermon was shared with Stephanie nineteen years later by Rachel Anderson, then a minister at the church.

feel *righteous anger* at injustice: Maximilian Agostini and Martijn Van Zomeren, "Toward a Comprehensive and Potentially Cross-Cultural Model of Why People Engage in

Collective Action: A Quantitative Research Synthesis of Four Motivations and Structural Constraints," *Psychological Bulletin* 147, no. 7 (2021): 667–700.

more likely to protest: Kenneth T. Andrews and Michael Biggs, "The Dynamics of Protest Diffusion: Movement Organizations, Social Networks, and News Media in the 1960 Sit-Ins," *American Sociological Review* 71, no. 5 (2006): 752–77; Michael Biggs, "Who Joined the Sit-Ins and Why: Southern Black Students in the Early 1960s," *Mobilization* 11, no. 3 (2006): 321–36.

more supportive of racial justice: Michael Biggs and Kenneth T. Andrews, "Protest Campaigns and Movement Success," *American Sociological Review* 80, no. 2 (2015): 416–43.

only 12 percent of Americans supported same-sex marriage: Shankar Vedantam and William Cox, "Hidden Brain: America's Changing Attitudes Toward Gay People," interview by Steve Inskeep, NPR, April 17, 2019.

one of the fastest-moving political issues in the nation's history: By 2021, 70 percent of Americans supported gay marriage; see Justin McCarthy, "Record-High 70% in U.S. Support Same-Sex Marriage," *Gallup News*, June 5, 2023.

peer permission to express their true beliefs: Leonardo Bursztyn, Alessandra L. González, and David Yanagizawa-Drott, *Misperceived Social Norms: Female Labor Force Participation in Saudi Arabia* (Chicago: Becker Friedman Institute for Research in Economics, 2018).

closed more than fifteen hundred polling places: Ed Pilkington and Jamie Corey, "Dark Money Groups Push Election Denialism on US State Officials," *Guardian*, April 5, 2023; Sam Levine and Kira Lerner, "Ten Years of a Crippled Voting Rights Act: How States Make It Harder to Vote," *Guardian*, June 25, 2023; "Democracy Diverted."

Democrats and Republicans oppose gerrymandering: Some polls put the numbers as high as 90 percent; here I'm citing the most conservative statistic. See Bryan Warner, "Polls Show Voters Nationwide and in NC Agree: Gerrymandering Must End," Common Cause North Carolina, April 7, 2021, https://www.commoncause.org/north-carolina/democracy-wire/polls-show-voters-nationwide-and-in-nc-agree-gerrymandering-must-end/; "Americans Are United Against Partisan Gerrymandering," Brennan Center for Justice, March 15, 2019, https://www.brennancenter.org/our-work/research-reports/americans-are-united-against-partisan-gerrymandering; John Kruzel, "American Voters Largely United Against Partisan Gerrymandering, Polling Shows," *The Hill*, August 4, 2021.

learned about Katie Fahey: Tina Rosenberg, "Putting the Voters in Charge of Fair Voting," *New York Times*, January 23, 2018.

cared about national issues: Katie Fahey, interview with author, September 21, 2023.

"take on gerrymandering in Michigan": Katie Fahey Schergala, "I'd like to Take on Gerrymandering in Michigan, If You're Interested in Doing This as Well, Please Let Me Know :)," Facebook, November 10, 2016, https://www.facebook.com/katie.rogala.3/posts/10153917724442633.

"They were everywhere": *Slay the Dragon*, directed by Chris Durrance and Barak Goodman (Participant, 2019).

districting committees like Michigan's: Nathaniel Rakich, "Did Redistricting Commissions Live Up to Their Promise?," FiveThirtyEight, January 24, 2022.

These questions have animated Loretta Ross: Loretta J. Ross, interviews with author, October 31, 2022, and November 23, 2022; additional quotes drawn from Loretta J. Ross, "Calling in the Calling Out Culture: Conversations Instead of Conflicts," *Critical Conversations Speaker Series* (Santa Monica, CA: New Roads School, 2021).

"the cannibalistic maw of cancel culture": Loretta J. Ross, "I'm a Black Feminist. I Think Call-Out Culture Is Toxic," *New York Times*, August 17, 2019.

Chapter 11: Our Common Fate

double the amount of warming agreed upon: Eric Roston, "Climate Projections Again Point to Dangerous 2.7C Rise by 2100," *Bloomberg News*, November 10, 2022.

billion-dollar natural disaster occurred once every three weeks: Nathan Rott, "Extreme Weather, Fueled by Climate Change, Cost the U.S. $165 Billion in 2022," NPR, January 10, 2023.

land that will be underwater by 2050: Denise Lu and Christopher Flavelle, "Rising Seas Will Erase More Cities by 2050, New Research Shows," *New York Times*, November 28, 2019.

"climate change is an unstoppable process": Adam Mayer and E. Keith Smith, "Unstoppable Climate Change? The Influence of Fatalistic Beliefs About Climate Change on Behavioural Change and Willingness to Pay Cross-Nationally," *Climate Policy* 19, no. 4 (2018): 511–23.

climate despair: "Climate Fatalism Grips Young People Worldwide While the Urgency for Solution-Oriented Media Grows," Ipsos, November 10, 2021.

most people think that most people don't care: Gregg Sparkman, Nathan Geiger, and Elke U. Weber, "Americans Experience a False Social Reality by Underestimating Popular Climate Policy Support by Nearly Half," *Nature Communications* 13, no. 1 (2022): 4779.

Hardin contracted polio: Biographical details drawn mainly from the Garrett Hardin Society's UCSB Oral History; see Garrett Hardin, "Garrett Hardin Oral History Project: Tape 1—'The Early Years,'" interview by David E. Russell, Garrett Hardin Society, May 22, 2005, https://www.garretthardinsociety.org/gh/gh_oral_history_tape1.html.

post-polio syndrome: John H. Tanton, "Garrett and Jane Hardin: A Personal Recollection," Garrett Hardin Society, October 29, 2003, https://www.garretthardinsociety.org/tributes/tr_tanton_2003oct.html.

"the one stable place in my life": Garrett Hardin, "Living in a World of Limits: An Interview with Noted Biologist Garrett Hardin," interview by Craig Straub, Garrett Hardin Society, June 9, 2003, https://www.garretthardinsociety.org/gh/gh_straub_interview.html.

numbers were exploding: Hardin, "Garrett Hardin Oral History Project: Tape 1."

"The Tragedy of the Commons": Garrett Hardin, "The Tragedy of the Commons: The Population Problem Has No Technical Solution; It Requires a Fundamental Extension in Morality," *Science* 162, no. 3859 (1968): 1243–48.

clash of two immutable laws: Hardin describes these laws as similar to thermodynamics, in Garrett Hardin, "Garrett Hardin Oral History Project: Tape 7," interview by David E. Russell, Garrett Hardin Society, June 9, 2003, https://www.garretthardinsociety.org/gh/gh_oral_history_tape7.html.

"comparable to the Underground Railway": Garrett Hardin, "Garrett Hardin Oral History Project: Tape 5—From the Lab to the Field of Ecology," interview by David E. Russell, Garrett Hardin Society, June 9, 2003, https://www.garretthardinsociety.org/gh/gh_oral_history_tape5.html.

"idea of a multi-ethnic society is a disaster": Garrett Hardin, "Lifeboat Ethics: The Case Against Helping the Poor," *Psychology Today* (1974); Matto Mildenberger, "The Tragedy of the *Tragedy of the Commons*," *Scientific American Blog Network*, April 23, 2019; Jason Oakes, "Garrett Hardin's Tragic Sense of Life," *Endeavour* 40, no. 4 (2016): 238–47.

population declining in twenty-five nations: Erik Nordman, *The Uncommon Knowledge of Elinor Ostrom: Essential Lessons for Collective Action* (Washington, DC: Island Press, 2021).

like a touring rock band: Garrett Hardin, "Garrett Hardin Oral History Project: Tape 10," interview by David E. Russell, Garrett Hardin Society, June 9, 2003, https://www.garretthardinsociety.org/gh/gh_oral_history_tape10.html.

today's doomers are less likely: Studies show that people are most likely to act when they feel anxiety about the climate *and* hope that the situation could improve. See Shanyong Wang et al., "Predicting Consumers' Intention to Adopt Hybrid Electric Vehicles: Using an Extended Version of the Theory of Planned Behavior Model," *Transportation* 43, no. 1 (2014): 123–43; Kimberly S. Wolske, Paul C. Stern, and Thomas Dietz, "Explaining Interest in Adopting Residential Solar Photovoltaic Systems in the United States: Toward an Integration of Behavioral Theories," *Energy Research & Social Science* 25 (2017): 134–51.

why bother cutting back: Roderick M. Kramer, "Trust and Distrust in Organizations: Emerging Perspectives, Enduring Questions," *Annual Review of Psychology* 50, no. 1 (1999): 569–98.

just ninety large companies: Douglas Starr, "Just 90 Companies Are to Blame for Most Climate Change, This 'Carbon Accountant' Says," *Science* 25 (2016).

top 1 percent of global earners produce twice as much pollution: Tim Gore, "Confronting Carbon Inequality: Putting Climate Justice at the Heart of the COVID-19 Recovery," *Oxfam International*, September 21, 2020.

pill powerful levers to uphold the status quo: Benjamin Franta, "Weaponizing Economics: Big Oil, Economic Consultants, and Climate Policy Delay," *Environmental Politics* 31, no. 4 (2021): 555–75.

"one of the most successful, deceptive PR campaigns maybe ever": Mark Kaufman, "The Carbon Footprint Sham: A 'Successful, Deceptive' PR Campaign," *Mashable*, July 13, 2020. See also Alvin Powell, "Tracing Big Oil's PR War to Delay Action on Climate Change," *Harvard Gazette*, September 28, 2021.

"carbon shamed": Aylin Woodward, "As Denying Climate Change Becomes Impossible, Fossil-Fuel Interests Pivot to 'Carbon Shaming,'" *Business Insider*, August 28, 2021.

options that Big Oil has lobbied against: Rebecca Solnit, "Big Oil Coined 'Carbon Footprints' to Blame Us for Their Greed. Keep Them on the Hook," *Guardian*, August 23, 2021.

deflects responsibility: Behavioral scientists Nic Chater and George Loewenstein think of this as emphasizing "i-frames," or individual causes, over "s-frames," or systemic causes; see Nick Chater and George Loewenstein, "The I-Frame and the S-Frame: How Focusing on Individual-Level Solutions Has Led Behavioral Public Policy Astray," *Behavioral and Brain Sciences* 46 (2022): e147.

Americans think: Sparkman, Geiger, and Weber, "Americans Experience a False Social Reality."

"Well, he, in my mind, became a totalitarian": Derek Wall, *Elinor Ostrom's Rules for Radicals: Cooperative Alternatives Beyond Markets and States* (Chicago: University of Chicago Press, 2017), 21–22.

"He's just made this up": Quotes by Elinor Ostrom at the New Frontiers in Global Justice Conference at UC San Diego; see diptherio, "Elinor Ostrom on the Myth of Tragedy of the Commons," August 9, 2014, https://www.youtube.com/watch?v=ybdvjvIH-1U.

"barefoot and pregnant": Biographical details about Ostrom drawn from Nordman, *Uncommon Knowledge of Elinor Ostrom*; Wall, *Elinor Ostrom's Rules for Radicals*.

"public entrepreneurship": Elinor Ostrom, "Public Entrepreneurship: A Case Study in Ground Water Basin Management" (PhD diss., University of California Los Angeles, 1964).

successful commons projects around the world: Nordman, *Uncommon Knowledge of Elinor Ostrom*; Elinor Ostrom and Harini Nagendra, "Insights on Linking Forests, Trees, and People from the Air, on the Ground, and in the Laboratory," *Proceedings of the National Academy of Sciences* 103, no. 51 (2006): 19224–31; Paul B. Trawick, "Successfully Governing the Commons: Principles of Social Organization in an Andean Irrigation System," *Human Ecology* 29, no. 1 (2001): 1–25.

"than he could feed during the winter": Elinor Ostrom, *Governing the Commons* (Cambridge: Cambridge University Press, 2015), 62.

"design principles": Elinor Ostrom, "Collective Action and the Evolution of Social Norms," *Journal of Economic Perspectives* 14, no. 3 (2000): 137–58.

"they'll cheat whenever they can": *Actual World, Possible Future*, directed by Barbara Allen (WTIU Documentaries, May 25, 2020), https://video.indianapublicmedia.org/video/actual-world-possible-future-09rkab/.

at least a billion people today sustainably govern themselves: *Actual World, Possible Future*, directed by Barbara Allen.

share their identity with neighbors, ancestors, and descendants: For more on this, see Marshall Sahlins, *The Western Illusion of Human Nature: With Reflections on the Long History of Hierarchy, Equality and the Sublimation of Anarchy in the West, and Comparative Notes on Other Conceptions of the Human Condition* (Chicago: Prickly Paradigm Press, 2008).

Renewables are expected to take over: David Gelles et al., "The Clean Energy Future Is Arriving Faster Than You Think," *New York Times*, August 17, 2023; International Energy Agency, "Renewables 2022: Analysis and Forecast to 2027," IEA, 2022, https://www.iea.org/reports/renewables-2022.

speeds the process while improving soil quality: James Dacey, "Sprinkling Basalt over Soil Could Remove Huge Amounts of Carbon Dioxide from the Atmosphere," *Physics World*, August 1, 2021.

rosy distraction from serious climate efforts: For a few examples of such criticism and controversy, see Nick Gottlieb, "The False Hope of Carbon Capture and Storage," *Canadian Dimension*, May 30, 2022; Robert F. Service, "U.S. Unveils Plans for Large Facilities to Capture Carbon Directly from Air," *Science Insider*, August 11, 2023; Genevieve Guenther, "Carbon Removal Isn't the Solution to Climate Change," *New Republic*, April 4, 2022.

as much pollution as adding two million gas vehicles to the roads: Ella Nilsen, "The Willow Project Has Been Approved: Here's What to Know About the Controversial Oil-Drilling Venture," CNN, March 14, 2023.

received over a million letters and thousands of calls: See Elise Joshi, "Please Watch Even If You've Seen the Original Video!," TikTok, September 7, 2023, https://www.tiktok.com/@elisejoshi/video/7276138179386985774?lang=en.

"Ok Doomer": Cara Buckley, "'OK Doomer' and the Climate Advocates Who Say It's Not Too Late," *New York Times*, June 22, 2023.

founder of Black Girl Environmentalist: Gatheru's videos can be found at "Wawa Gatheru (@wawagatheru)," TikTok, n.d., https://www.tiktok.com/@wawagatheru/.

"Fear doesn't motivate people towards sustainable action": David Gelles, "With TikTok and Lawsuits, Gen Z Takes on Climate Change," *New York Times*, August 21, 2023.

"compensation pledges": "COP27 Climate Summit: Here's What Happened on Tuesday at the COP27 Climate Summit," *New York Times*, November 9, 2022.

"clean and healthful environment": David Gelles and Mike Baker, "Judge Rules in Favor of Montana Youths in a Landmark Climate Case," *New York Times*, August 16, 2023.

barred oil drilling: Timothy Puko, "Biden to Block Oil Drilling in 'Irreplaceable' Alaskan Wildlands," *Washington Post*, September 7, 2023.

"That's why we're here": Annenberg School for Communication, "Emile."

Epilogue

the family is thriving: Stephanie Bruneau, interview with author, September 1, 2023.

Appendix A

The monk and author Pema Chödrön writes: Pema Chödrön, *Taking the Leap: Freeing Ourselves from Old Habits and Fears*, ed. Sandy Boucher (Boston, MA: Shambhala, 2010).

Index

accountability, for carbon emissions, 195–196
activism. *See* social activism
affirmation of values, 27–28
Aiken, Philip, 196–197
Ainsworth, Mary, 36–37
Alaska oil drilling, 196, 197
Amazon, 129
American Psychological Association, 168
Andropov, Yuri, 130–131
anomie, 55
anti-rape activism, 181–182
Arendt, Hannah, 171
Aristotle, 162
asset-framing, 73–78, 123–124, 166, 176–177, 197, 211
"at-risk" youth, 73
attachment styles, 36–43
   insecure attachment, 36–38
   safe home base for secure attachment, 38–43
   theory of, 36–37
autarkeia (self-sufficiency), 16, 23, 25–28
Axelrod, Robert, 84–86

Ballmer, Steve, 117–118, 119
Bangladesh, 75
Beckett, Samuel, 169
behavioral experiments, in cognitive behavioral therapy, 44–45
beliefs and values
   affirmation of values, 27–28
   of Emile, 3, 4, 8–9, 23–24, 25–28
   primal beliefs, 62–63
   study of, 26–27
Black Girl Environmentalist, 197
Black Panther Party, 112
Bohns, Vanessa, 81
Boivert, Franck, 89–90
Bornstein, David, 69, 71, 75–77, 176
Boston firefighters, 79–80, 82–83, 121
Boston Globe, 79–80, 83
Bradley, Bill, 120
Brazil fishing communities, 46–48
British Petroleum (BP), 190
Browning, Elizabeth Barrett, 194
Bruneau, Emile
   on activism, 172–173
   background, 1–2, 8
   beliefs and values of, 3, 4, 8–9, 23–24, 25–28
   childhood and early education, 2–3, 23–24, 25, 38–39, 55–58, 62–63, 88–89
   Colombian peace research by, 139–143
   on cross-party conversations, 145
   health issues and death of, 3, 9, 40–41
   as hopeful skeptic, 33
   immigration study by, 133–134
   legacy of, 9, 58, 200–202
   organizational leadership style of, 127
   Peace and Conflict Neuroscience Lab, 1, 25, 58, 127, 139–143
   peace work by, 1, 24–25, 58

Bruneau, Emile *(cont.)*
   on power of community, 199
   on racism, 87
   as rugby player and coach, 89–91
   skepticism mindset of, 39–40
   wife and children of, 3, 8–9, 40–41, 201–203
Bryant, Fred, 66
burnout, 103–104
A Burst of Light (Lorde), 112
Byrne, David, 76, 77–78

California
   Peninsula School, 55–58, 62–63, 89, 201
   preexisting conditions for cynicism in, 50–51, 64
   water supply problems in, 191–192
callouts, in activism, 182–183
Canada, mistrust experiment in, 65–66
Carbon Engineering, 195
carbon footprint, 189–190, 193–196
carbon removal, 195–196
Carlin, George, 38
Carnegie, Andrew, 120, 151
Casas, Andrés, 137–143
Casas, Juan, 140–142
cash transfer programs, 163–165
CBT (cognitive behavioral therapy), 44–45
change, neuroscience of, 87
charities (philanthropy), 21, 51, 65–66, 72–73, 164–166
Charter 77, 170, 171–172
ChatGPT, 125
cheaters
   detection of, 65–67, 79–80
   prisoner's dilemma game and, 84–86
   transformation of, 87–88
childhood poverty, 156, 158–159
Chödrön, Pema, 209
Christian Social Gospel, 153
Chung Sye-kyun, 23
Citizens United, 134, 161
Civic Forum, 172
Civil Rights movement, 174

claim ratings, 213–224
   about, 213–215
   by chapter, 216–224
classism, 151–152
classroom hierarchy, 126
Clifton, Jer, 62
climate action, 185–199
   background, 185–186
   choosing a future, 193–199
   tragic view of the commons, 186–191
   victory of the commons, 191–193
Clínica de la Raza, 64
cognitive behavioral therapy (CBT), 44–45
Cohen, Geoff, 26–27, 28
collective action, 172–175
Colombia, 137–143
commodification, 51–55, 58–59
Common Ground, 162
communal relationships
   building of, 21–23, 40, 64
   characteristics of, 46–48, 51–54, 58–59, 199
   mutual aid communities, 112–113
   well-being and, 20–21
Communism, 49, 137–142, 156, 169–172, 175
compensation pledges, in climate crisis, 197
conflict entrepreneurs, 135–137
ConocoPhillips, 196
conspiracy theories, 26, 32, 33–36, 41, 42
constitution of knaves, 155–158, 160–161, 162, 166
Cook, Walter, 18, 20, 27
cooperation
   as inborn, 65, 112, 200
   misperceptions about, 65–67
   in Ocean Villages versus Lake Towns, 46–48, 54–55, 58–64, 97–98, 121, 192
   prisoner's dilemma game, 84–86
   public goods problem and, 68
   trust and, 21–22, 23, 40, 64, 86
   underbearing attentiveness for, 128
Cornet, Manu, 114, 115
cosmopolitanism, 16, 23

Index

counterintelligence work, 83–84, 86
counting and uncounting, 58–61, 91, 108–110, 211
COVID pandemic
  author's worldview during, 3–4
  community trust-building during, 22–23
  conspiracy theories, 33, 35
  elite abuse during, 49–51, 64
  Microsoft and, 124–125
  mutual aid programs during, 113
  negativity bias and, 66
  organizational cynicism during, 118–119, 129
  social prescriptions during, 107
  welfare programs during, 155
creative maladjustment, 168, 171–172, 181–183
crime waves, 70–71, 74
Cuban Missile Crisis, 92, 144
cultures of trust, 114–129. *See also* organizational cultures
cynical genius illusion, 29–31, 33
cynicism
  alternatives to, 46–48, 54–64
  ancient Cynicism, 15–16
  ancient Cynicism principles, 16–17, 23–28
  defined, 9
  effects of, 4–6, 19–20 (*See also* organizational cultures; politics and political parties; social interactions)
  elite abuse and, 49–51, 54–55, 64, 172 (*See also* inequality)
  future without cynicism, 9–11, 200–203 (*See also* climate action; hopeful skepticism; social activism)
  myths of, 6–7
  as status quo tool, 7, 146, 166, 170–171, 189
  stereotypes of cynics, 6, 29–31, 33
  theory of, 17–20
  transactional behavior, 51–54, 58–59
  unlearning, 7–9 (*See also* Bruneau, Emile; freedom from cynicism; media)
  Cynics on, 15–17
Cyprus, 137
Czechoslovakia, 169–170, 171–172, 175

Darwin, Charles, 116. *See also* social Darwinism
deficit-framing, 73–74, 75
democratic norms, 49
Democrats. *See* politics and political parties
Desmond, Matthew, 160
Diogenes Club, 15–16
Diogenes of Sinope, 16–17, 23, 28
direct cash transfer programs, 163–165
disabilities, 157–158
disagree better, 144–146, 212
donations, 21, 51, 65–66, 72–73
Douglas, Karen, 35–36
Dreeke, Robin, 83–84, 86
Durkheim, Émile, 54–55

earned attachment, 41
earned trust, 82
earthquake, in Japan (1995), 21–22, 23, 40, 64
East Germany, 49, 156
economicus, 115–119
Edelman's "Trust Barometer," 120
efficacy toward injustice, 173–175
elite abuse, 49–51, 54–55, 64, 172. *See also* inequality
Emerson, Ralph Waldo, 26, 28, 202
Emile. *See* Bruneau, Emile
empathy, 35, 40, 102, 145, 146–147
encounter counting experiment, 108–110, 211
evidence evaluation
  about, 213–215
  chapter claim ratings, 217–224
exchange relationships, 51–54

Facebook, 53, 61, 177–179
fact-checking, 210–211
Fahey, Katie, 176–180

Index

faith
  in humanity, 5
  leaps of, 88–93, 162–167
The Fall of the Cabal (video), 35
false polarization, 131–135, 136
FARC (rebel army), 137–142
firefighters, 79–80, 82–83, 121
529 college savings program, 160–161
Fixes (newspaper column), 76–77
food stamps program, 154–155, 159–160, 162
Foundations for Social Change, 164–166
freedom from cynicism, 79–93
  leaps of faith and, 88–93
  and preemptive strikes, 79–80, 82–83, 86, 91
  reciprocity mindset for, 87–88, 91, 212
  and self-fulfilling prophecies, 80–83, 93
  trust as power, 83–86
Fridays for Future movement, 196
Friedman, Milton, 164
Fruitvale neighborhood, 50–51, 64
Fuller, William, 181–182

Gaetz, Matt, 155
Gatheru, Wanjiku, 197, 198
General Electric (GE), 115, 117, 121
General Social Survey (GSS), 4–5, 153
Generous Tit for Tat (GTFT) program, 85–86, 92–93
Gerena, Jesús, 163
Germanic tribes, 134–135
Germany, 49, 156
gerrymandering, 176–180
"good true self" effect, 73–74
Goodwin, William, 156–158, 160, 162–163
gossip, 67–73, 78
Gould, Jay, 151
government. *See also* politics and political parties
  cynics' view of, 6
  elite abuse and, 49–51, 64
  social programs, 154–155, 157–161, 162
Grameen Bank, 75

GSS (General Social Survey), 4–5, 153
GTFT (Generous Tit for Tat) program, 85–86, 92–93

Hardin, Garrett, 187–189, 191, 193
Harris, Jen, 126
hatred, as brain disease, 25
Havel, Václav, 169–170, 171–172, 175
Hearst, William Randolph, 152
hell. *See* media
Hemlock Society, 187
high-trust communities, 20–23
hikikomori, 97–98, 101, 105–106, 110–112
Hobbes, Thomas, 6
Hogan, Kathleen, 124–125
homo collaboratus, 123–126, 127
homo economicus, 115–119, 121–122, 127, 156, 162
hope. *See also* hopeful skepticism
  of Cynics, 23–25
  mythological stories on, 7–8
  optimism versus, 8
  for politics, 146–147
  as skill, 4
hopeful skepticism, 29–45
  background, 9–11
  cynical genius illusion, 29–31, 33
  as cynicism antidote, 41–45
  defined, 9
  development and guide to, 38–43
  development of and guide to, 209–212
  for disappointed idealists, 33–38
  leaps of faith and, 91–93
  for media consumption, 74, 77–78, 211
  as scientific mindset, 31–33
hopelessness
  climate crisis and, 194–195
  political, 136–137
  as privilege, 196–198
The HP Way (Packard), 121
humanity
  faith in, 5
  love of, 16–17, 23

Index

Hume, David, 155–156
Hungary, 130

Iglesias, Chris, 64
Immelt, Jeff, 117
immigration, 50, 133–134
improv, 179
inequality, 151–167
    background, 48, 151–154
    constitution of knaves, 155–158, 160–161, 162, 166
    elite abuse, 49–51, 54–55, 64, 172
    framing of, 73
    poverty and, 158–159
    redistributing trust, 162–167
    stereotypes and tropes, 154–155, 157–158
    trust for the few, 158–162
Inflation Reduction Act (2022), 197
influence neglect, 81
"investment" game, 19, 47, 81–82, 88
Israel, 137

January 6 insurrection, 131
Japan
    community trust-building example in, 21–22, 23, 40, 64
    hikikomori, 97–98, 101, 105–106, 110–112
Jesus, 17
Jim Crow policies, 152
Johnson, Boris, 49–50
Jude, Frank, Jr., 50

Kennedy, John F., 92, 144
Keynes, John Neville, 115–116
Khrushchev, Nikita, 92, 144
King, Martin Luther, Jr., 168
kintsugi, 111
knaves, 155–158, 160–161, 162, 166
Kobe neighborhoods of Mikura and Mano (Japan), 21–22, 23, 40, 64
kosmopolitês, 16, 23

Kropotkin, Peter, 112, 120
Kteily, Nour, 145
Ku Klux Klan, 183

Lake Towns versus Ocean Villages, 46–48, 54–55, 58–64, 97–98, 121, 192
Lane, Logan, 52, 59–60
Lazarus, Richard, 8
leadership in organizations, 123–129
leaps of faith, 88–93, 162–167
Leibbrandt, Andreas, 46–47
Lembke, Anna, 52
Leviathan (Hobbes), 6
Lewis, Janet, 90–91
LGBTQ activism, 174
Lincoln Middle School, 122–123, 125–126, 127, 183
local trust, 63–64
Lorde, Audre, 112
loud trust, 91–93, 128–129, 144, 163, 212
Lowrey, Annie, 159
low-trust communities, 20–23
Luddite Club, 60
lynching, 152

"male guardianship" laws, 175
Manchin, Joe, 155
Mandela, Nelson, 175
Mano (Japan) neighborhood, 21–22, 23, 40, 64
market creep, 52, 54–55
McKibben, Bill, 199
Medellín Miracle, 138–139
media, 65–78
    alternatives to negativity bias, 73–78
    gossip and, 67–73, 78
    on grassroots campaigning, 178
    guide for transforming cynicism to skepticism in, 74, 77–78, 211
    negativity bias of, 65–67
    post–U.S. Civil War, 151–152
Medley, Donald, 18, 20, 27
Michigan, gerrymandering in, 176–180

microloans, 75
micromanaging workers, 118–119, 120, 121–122
Microsoft, 114, 115, 117–118, 119, 121, 123–125, 127
Mikura (Japan) neighborhood, 21–22, 23, 40
mind freeze, 138–143
Moore-Berg, Samantha, 58
morality, of cynicism, 7
moral responsibility, 153
Murthy, Vivek (surgeon general), 99, 107
Mutual Aid (Kropotkin), 112
mutual aid communities, 112–113

Nadella, Satya, 123–124, 127
Naikan, 105–106
naive optimism, 6–7
Nance, Malcolm, 135–136
National Health Service (NHS), 107
natural disasters, 21–22, 40
negative income tax, 164
negativity bias
  alternatives to, 73–78
  in media, 8, 65–73
  in politics, 133, 135, 142
  preemptive strikes of, 79–80, 82–83
  self-fulfilling prophecies of, 80–83
  small talk and, 100–101
Neuropaz, 143
news outlets. *See* media
New York middle school, 122–123, 125–126, 127, 183
New York Times, 76–77
NHS (National Health Service), 107
Nietzsche, Friedrich, 8
"No" campaign, in Colombia, 140–141
nut picking, 136, 154–155

Oakland (California), 50–51, 64
Oath Keepers, 131, 135
Ocean Villages versus Lake Towns, 46–48, 54–55, 58–64, 97–98, 121, 192
oil drilling, in Alaska, 196, 197

online cynicism, 121–123
online dating, 53–54
OpenAI, 125
Operation RYaN, 130–131, 135
oppression, 146–147
optimism. *See also* social activism
  false security of, 78
  hope versus, 8
  of idealists, 34–35
  naive optimism, 6–7
  theory of, 18
organizational cultures, 114–129
  background, 114
  growth of cynicism in, 121–123
  homo collaboratus and, 123–126
  homo economicus and, 115–119, 121–122, 127
  price of cynicism in, 119–121
  trust leadership, 123–129
Ostrom, Elinor, 191–193
"other care," 110–113

Packard, David, 121
Pandora, 7–8
parenting and parenthood
  attachment styles and, 36–43
  hikikomori and, 98, 101, 111
  as life-changing, 28, 38–39
  safety and trust rules, 61–63
  with underbearing attentiveness, 24, 39–41
Paris Climate Accords (2015), 186, 193–194
"Partygate" scandal (2021), 49–50
Peace and Conflict Neuroscience Lab, 1, 25, 58, 127, 139–143
Peninsula School, 55–58, 62–63, 89, 201
The People (organization), 180
Petrova, Kate, 213–215
philanthropia (love of humanity), 16–17, 23
philanthropy (charities), 21, 51, 65–66, 72–73, 164–166
Pinker, Steven, 54

Plessy v. Ferguson, 152
polarization (false), 131–135, 136
politics and political parties, 130–147
    background, 130–131
    beyond peace to hope, 146–147
    conflict entrepreneurs, 135–136
    cross-party conversations, 144–146
    disagreeing better, 145–146
    elite abuse and, 49–50
    false polarization, 131–135, 136
    misperceptions, unwinding of, 7, 137–143
    political hopelessness, 136–137
    post–U.S. Civil War, 151–152
    voter suppression, 176–180
"positive deviants," 76
poverty
    education and, 123
    inequality and, 151–152, 154–155, 157–161
    media coverage of, 75–76
    mutual aid programs and, 112–113
    redistributing trust and, 162–167
Poverty, By America (Desmond), 160
povertyism, 159
The Power of the Powerless (Havel), 169–170, 171–172
Prague, 169–170, 171–172, 175
Prague Spring, 169–170
pre-disappointment, 38, 86
preexisting conditions for cynicism, 46–64
    alternative example, 46–48, 54–58
    alternatives, manual for, 58–64
    commodification, 51–55, 58–59
    elite abuse, 49–51, 54–55, 64
    inequality, 48, 54–55
    primal beliefs, 62–63
Prisoners Against Rape, 182, 183
prisoner's dilemma game, 84–86
privilege of hopelessness, 196–198
profit sharing, 115
progressive movement, 152–154
Prometheus, 7
propaganda, 170–171

public entrepreneurship, 192
public goods game, 68
punishment, culture of, 122–123, 125–126, 127
punk rock, 138
Putin, Vladimir, 170–171
Putnam, Robert, 151, 153

QAnon conspiracy theory, 33–36, 41, 42

race and racism
    in education, 73, 123
    inequality and, 152, 155, 157–158
    mutual aid programs as protest against, 112
    neuroscience of, 87
    power and, 146
    as preexisting condition for cynicism, 50–51
    social activism and, 174, 181–182
    tropes and stereotypes, 155
    voter suppression and, 152
RAND Corporation, 170–171
"rank and yank," 117, 119, 120, 124
Reagan, Ronald, 154
reality testing, 44–45, 106–110
Reasons to Be Cheerful (magazine), 76, 77–78
reciprocity mindset, 87–88, 91, 212
Republicans. *See* politics and political parties
restorative justice, 125–126, 183
Rhodes, Stewart, 131, 135–136
righteous anger, 173–175
Ripley, Amanda, 135–136
robber barons, 151–152
Rockefeller, John D., 116
Roman Empire, 134–135
Rosenberg, Tina, 76–77, 176
Ross, Loretta, 181–183, 184
rugby, 89–91
Russia. *See also* USSR
    propaganda, 170–171
    spying by, 84

same-sex marriage, 174
Sanders, Bernie, 35
Santos, Juan Manuel, 139
Saudi Arabia, 175
savoring practices, 63
Schaaf, Libby, 50
Schultze, Charles, 51
self-care, 103–104, 111–113
self-fulfilling prophecies, 80–83
Self-Reliance (Emerson), 26
self-sufficiency, 16, 23, 25–28
sexist tropes, 154–155
Shaw, George Bernard, 30
Shorters, Trabian, 72–73
signs and symptoms of cynicism, 15–28
   ancient Cynicism, 15–16
   ancient Cynicism principles, 16–17, 23–28
   background, 9, 15
   theory of cynicism, 17–20
   trust gaps as, 17–23
Sisyphus, 173
sit-ins, 174
SJN (Solutions Journalism Network), 77
skepticism mindset. *See* hopeful skepticism
social activism, 168–184
   background, 7, 168–169
   community trust-building through, 21–23, 40, 64
   emotional alloy of change, 172–175
   everyone's miracles, 175–180
   growing the tent, 181–184
   impossibility and, 169–172
social Darwinism, 116–118, 120, 151–153, 161–162
Social Gospel, 153
social health, diseases of, 9, 15. *See also* preexisting conditions for cynicism; signs and symptoms of cynicism
social interactions, 97–113
   for burnout and self-care, 103–105, 111–113
   cynicism and, 99–103, 106
   health effects of, 98–99, 107
   isolation and *hikikomori*, 97–98, 101, 105–106, 110–112
   misperceptions and reality testing, 105–110
   myths of, 6–7, 65–67
   of non-cynics, 20, 211
   "other care," 110–113
   uncounting of, 59–61, 91, 109–110
social media, 2, 52–53, 59–61, 70, 136, 177–179
social prescriptions, 107
social savoring, 63
social wisdom, 44–45
solutions journalism, 76–77
Solutions Journalism Network (SJN), 77
Solutions Story Tracker, 77, 78, 176–177, 211
South Korea, 22–23
spies, 83–84, 86
Stanford, Leland, 151
Stanford rugby teams, 89–91
Starbucks, 129
Stasi, 49, 156
status quo
   replacing, 174–175
   tool of cynicism, 7, 146, 166, 170–171, 189
stereotypes and tropes
   of cynics, 6, 29–31, 33
   of inequality, 154–155, 157–158
Supplemental Nutrition Assistance Program (SNAP), 154–155
surveillance of workers, 118–119, 120, 121–122
sustainable living principles, 191–193, 196–197. *See also* climate action
Syracuse middle school, 122–123, 125–126, 127, 183

task interdependence, 124
tax breaks, 160–161
Taylor, Linda, 154
teachers and teaching philosophies
   cynicism and, 18, 122–123, 125

of Peninsula School, 55–58,
    62–63, 89, 201
  racism and, 73, 123
  restorative justice and, 125–126, 183
  trust and, 56–58, 127
Temporary Assistance for Needy Families
    program, 155
Teo, Alan, 101
They Want to Kill Americans (Nance), 135
Thunberg, Greta, 196
Tinder (app), 53–54
Tit for Tat program, 85–86
Toronto Star, 65–66
transactional behavior, 51–55, 58–59
tropes. *See* stereotypes and tropes
trust
  abuse of, 48–51 (*See also* inequality)
  attachment styles and, 36–43
  gaps in, 17–23, 25–28, 40, 63–64
  investment game, 19, 47, 81–82, 88
  leaps of faith and, 91–93
  local trust and neighborhoods, 21–22,
    23, 40, 63–64
  loud trust, 91–93, 128–129, 144, 163, 212
  myths of, 6, 7
  naive trusters, 31, 32–33
  in Ocean Villages vs. Lake Towns, 46–48,
    54–55, 58–64, 97–98, 121, 192
  in organizations, 123–129 (*See also*
    organizational cultures)
  parenting safety rules and, 61–63
  as power, 86
  as skill, 4
  stereotypes and, 30
  turning to, as default behavior, 54–58
  underbearing attentiveness and, 24,
    39–41, 56–58, 90–91, 128–129
trust loudly, 91–93, 128–129, 144, 163, 212
Turkel, Marc, 117–118, 119

Ubuntu, 193
uncounting, 58–61, 91, 109–110
underbearing attentiveness, 24, 39–41,
    56–58, 90–91, 128–129

union membership, 129
United Auto Workers, 115, 129
United Kingdom
  cynicism in, 49–50
  social prescriptions in, 107
United Nations Climate Change
    Conference, 197
Unity Council, 64
UpTogether, 162–164, 166
USSR. *See also* Russia
  Cuban Missile Crisis, 92–93
  Operation RYaN, 130–131, 135

values. *See* beliefs and values
Velvet Revolution, 172
"Violent or Disruptive Incident
    Reporting," 123, 126
Visigoths, 134–135
Vonnegut, Kurt, 20, 79, 169
Voters Not Politicians, 178–180
voter suppression, 78, 146–147, 152, 176–180
Voting Rights Act (1965), 176

Walker, Gabrielle, 194–196
Wall Street (film), 116
Watanabe, Atsushi, 97–98, 101, 105–106,
    110–112
water supply problems, 191–192
Welch, Jack, 115, 117, 118, 127, 156
welfare queen myths, 154–155
"well-adjusted," 168
Wells Fargo, 121
White, LaJuan, 122–123, 125–126,
    127, 183
Willow Project, 196
Wilson, Woodrow, 152
wisdom of hope. *See* hopeful skepticism
workplace cynicism, 46–48, 114–123, 127.
    See also organizational cultures

Yousafzai, Malala, 175
Yunus, Muhammad, 75

zero-sum mentality, 48, 56, 119, 124, 151